# FERAL ANIMALS IN THE AMERICAN SOUTH

*An Evolutionary History*

The relationship between humans and domestic animals has changed in dramatic ways over the ages, and those transitions have had profound consequences for all parties involved. As societies evolve, the selective pressures that shape domestic populations also change. Some animals retain close relationships with humans, but many do not. Those who establish residency in the wild, free from direct human control, are technically neither domestic nor wild: they are feral. If we really want to understand humanity's complex relationship with domestic animals, then we cannot simply ignore the ones who went feral. This is especially true in the American South, where social and cultural norms have facilitated and sustained large populations of feral animals for hundreds of years. *Feral Animals in the American South* retells southern history from this new perspective of feral animals.

Abraham H. Gibson is a Fellow in Residence at the Consortium for History of Science, Technology and Medicine. He also teaches in the Department of History and Sociology of Science at the University of Pennsylvania. He has published extensively and has earned fellowships from the National Science Foundation and the Smithsonian Institution.

**Studies in Environment and History**

Editors

J. R. McNeill, *Georgetown University*
Edmund P. Russell, *University of Kansas*

Editors Emeritus

Alfred W. Crosby, *University of Texas at Austin*
Donald Worster, *University of Kansas*

Other Books in the Series

Ling Zhang *The River, the Plain, and the State: An Environmental Drama in Northern Song China, 1048–1128*
Andy Bruno *The Nature of Soviet Power: An Arctic Environmental History*
David A. Bello *Across Forest, Steppe, and Mountain: Environment, Identity, and Empire in Qing China's Borderlands*
Erik Loomis *Empire of Timber: Labor Unions and the Pacific Northwest Forests*
Peter Thorsheim *Waste into Weapons: Recycling in Britain during the Second World War*
Kieko Matteson *Forests in Revolutionary France: Conservation, Community, and Conflict, 1669–1848*
George Colpitts *Pemmican Empire: Food, Trade, and the Last Bison Hunts in the North American Plains, 1780–1882*
Micah Muscolino *The Ecology of War in China: Henan Province, the Yellow River, and Beyond, 1938–1950*
John Brooke *Climate Change and the Course of Global History: A Rough Journey*
Emmanuel Kreike *Environmental Infrastructure in African History: Examining the Myth of Natural Resource Management*
Paul Josephson, Nicolai Dronin, Ruben Mnatsakanian, Aleh Cherp, Dmitry Efremenko, and Vladislav Larin *An Environental History of Russia*
Gregory T. Cushman *Guano and the Opening of the Pacific World: A Global Ecological History*
Sam White *Climate of Rebellion in the Early Modern Ottoman Empire*
Alan Mikhail *Nature and Empire in Ottoman Egypt: An Environmental History*
Edmund Russell *Evolutionary History: Uniting History and Biology to Understand Life on Earth*
Richard W. Judd *The Untilled Garden: Natural History and the Spirit of Conservation in America, 1740–1840*
James L. A. Webb, Jr. *Humanity's Burden: A Global History of Malaria*
Frank Uekoetter *The Green and the Brown: A History of Conservation in Nazi Germany*
Myrna I. Santiago *The Ecology of Oil: Environment, Labor, and the Mexican Revolution, 1900–1938*

(*continued after Index*)

# FERAL ANIMALS IN THE AMERICAN SOUTH

An Evolutionary History

Abraham Gibson

*Consortium for History of Science, Technology, and Medicine*

# CAMBRIDGE
UNIVERSITY PRESS

University Printing House, Cambridge CB2 8BS, United Kingdom

One Liberty Plaza, 20th Floor, New York, NY 10006, USA

477 Williamstown Road, Port Melbourne, VIC 3207, Australia

314-321, 3rd Floor, Plot 3, Splendor Forum, Jasola District Centre, New Delhi - 110025, India

79 Anson Road, #06-04/06, Singapore 079906

Cambridge University Press is part of the University of Cambridge.

It furthers the University's mission by disseminating knowledge in the pursuit of education, learning and research at the highest international levels of excellence.

www.cambridge.org
Information on this title: www.cambridge.org/9781316610091

© Abraham Gibson 2016

This publication is in copyright. Subject to statutory exception and to the provisions of relevant collective licensing agreements, no reproduction of any part may take place without the written permission of Cambridge University Press.

First published 2016
First paperback edition 2018

*A catalogue record for this publication is available from the British Library*

*Library of Congress Cataloging in Publication data*
Names: Gibson, Abraham, 1983-, author.
Title: Feral animals in the American South : an evolutionary history / Abraham Gibson, Consortium for History of Science, Technology, and Medicine.
Description: New York, NY : Cambridge University Press, 2016. |
Series: Studies in environment and history
Identifiers: LCCN 2016013387 | ISBN 9781107156944 (hardback)
Subjects: LCSH: Feral animals–South Atlantic States–History. |
Domestic animals–South Atlantic States–History. |
Human-animal relationships–South Atlantic States–History. |
Coevolution–South Atlantic States–History.
Classification: LCC SF140.F47 G53 2016 | DDC 636.00975–dc23
LC record available at https://lccn.loc.gov/2016013387

ISBN 978-1-107-15694-4 Hardback
ISBN 978-1-316-61009-1 Paperback

Cambridge University Press has no responsibility for the persistence or accuracy of URLs for external or third-party internet websites referred to in this publication, and does not guarantee that any content on such websites is, or will remain, accurate or appropriate.

*Dedicated to my parents,*

*WILLIAM BENTLEY GIBSON*
*&*
*MARTHA SUE WOLFE GIBSON*

# EPIGRAPH

**fe·ral**

/ˈfi(ə)rəl/

Adjective; Latin – *feralis* (funereal; of or relating to the dead)

1 – Having escaped from domestication and returned to the wild

2 – Wild; savage; ferocious

# CONTENTS

| | |
|---|---|
| *List of Figures and Maps* | *page* x |
| *Preface* | xi |
| *Acknowledgments* | xii |

1. The trouble with ferality: domestication as coevolution and the nature of broken symbioses — 1
2. Making and breaking acquaintances: the origins of wildness, domestication, and ferality in prehistoric Eurasia — 13
3. When ferality reigned: establishing an open range in the colonial South — 27
4. Nascent domestication initiatives and their effects on ferality: claiming dominion in the antebellum South — 52
5. Anthropogenic improvement and assaults on ferality: divergent fates in the industrializing South — 80
6. Everything in its right place: wild, domestic, and feral populations in the modern South — 113

    Epilogue. Cultivating ferality in the Anthropocene: lessons for the American South and beyond — 139

| | |
|---|---|
| *Notes* | 143 |
| *Index* | 223 |

# FIGURES AND MAPS

| | | |
|---|---|---|
| 1 | Map of southeastern North America | page 7 |
| 2 | John White's watercolor of an Indian village, 1585 | 31 |
| 3 | William Bartram's sketch of the Great Alachua Savanna, 1765 | 47 |
| 4 | Secretary of War honors dogs who served in Second Seminole War, 1840 | 54 |
| 5 | Dog chasing a runaway slave, 1855 | 55 |
| 6 | George Washington's advertisement for equine stud services, 1787 | 61 |
| 7 | Enslaved people and their animals in the South Carolina Lowcountry, 1856 | 68 |
| 8 | Pigs being driven to railroad cars, 1857 | 73 |
| 9 | Horse statue outside the Virginia Historical Society in Richmond, Virginia | 77 |
| 10 | Maps showing distribution of pigs and corn, 1929–1930 | 82 |
| 11 | Cyrus Martin and feral pig, Ossabaw Island | 83 |
| 12 | Stock Law map of North Carolina, 1902 | 86 |
| 13 | The razorback hog, 1887 | 89 |
| 14 | Escorting feral horses from Chincoteague back to Assateague, 1887 | 94 |
| 15 | Atlanta Kennel Club's first dog show, 1900 | 105 |
| 16 | Mural depicting fabled origin of Assateague Horses | 118 |
| 17 | Distribution of pigs in the South, 2012 | 122 |
| 18 | Map showing range expansion of feral pigs, wild boars, and hybrids | 126 |
| 19 | Jamison Stone and "Monster Pig," 2007 | 129 |
| 20 | Photograph of trailer at Miami-Dade Animal Services | 134 |

# PREFACE

Historians once considered the "history of animals" such an unlikely subject that they openly ridiculed the idea in peer-reviewed journals, but that is no longer the case.[1] Over the past several decades, historians have published countless academic books and scholarly articles on the topic, and, in the words of Harriet Ritvo, the profession is currently experiencing an "animal turn."[2] Despite this growing scholarly interest, however, years of research have convinced me that the general public still regards the history of animals as an unfamiliar topic. That being the case, many readers might find the following information helpful before diving in. First, humans are obviously animals, but, in the interest of narrative thrust, all references to "animals" within this text refer to nonhuman animals. Also, plural first-person pronouns like "we" and "our" are reserved for all humans and only all humans. Within the context of this study, these pronouns do not include animals, and they are never used in reference to some people but not others. Finally, the reader will notice many instances in which the narrative references "animals who" rather than "animals that." This subtle distinction recognizes that animals are not lifeless automata, and that they are instead sentient organisms whose actions have influenced the course of life on Earth.[3]

# ACKNOWLEDGMENTS

If this is the only part of the book anyone ever reads, I will not complain. Many different people have helped me on this project over the years, and I hope to err on the side of effusion when expressing my gratitude. I will begin by acknowledging my friend and former graduate advisor at Florida State University, Fritz Davis, who has helped shepherd this project from its very beginning. His contributions to the project are significant. I'd also like to thank Ron Doel, Andrew Frank, Kristine Harper, James Jones, and Jennifer Koslow in the Department of History at Florida State University, as well as Michael Ruse and the History and Philosophy of Science program at FSU. While I am thanking educators who have influenced me, I also want to acknowledge previous teachers at other institutions, including Mark Barrow, Richard Burian, Tom Ewing, Matthew Goodrum, and Amy Nelson at Virginia Tech, Lucy Corin at UC Davis, Darlin' Neal at the University of Central Florida, William R. Alexander at the University of Mary Washington, R. E. Burnett at the National Defense University, Helen Storey at Penn State University, Edwina Prunty and Bobby Thompson at Ferrum College, and Mark Facknitz, Judy Good, Rose Gray, and Clive Hallman at James Madison University. I also want to thank *everyone* who has ever been associated in any way with the public school system in Franklin County, Virginia, and I am especially grateful toward my former English teachers, Terri Robertson, Raymond Williams, Jane Warren, Carla Tyree, and Leanne Worley.

I am thrilled that I can also finally thank the various people who have funded this project over the years. I especially want to thank Daniel and Sylvia Walbolt for endowing a five-year fellowship in the Department of History at Florida State University. This fellowship allowed me to study the mysteries of life for a living, and I sincerely thank them for their extraordinary generosity. I am also indebted to the Smithsonian Institution, which provided me with a Predoctoral Fellowship to support research on this project. In particular, I want to thank Pamela Henson at the Smithsonian Institution Archives and Jeffrey Stine at the National Museum of American History for their invaluable assistance on this project during my tenure in DC. While I am at it, I want to thank Ellen Alers, Courtney Bellizzi, Kira Cherrix, Brian Daniels, Greg Palumbo, Blake Scott, Mitch Toda, Anne Van De Camp, and everyone else at the Smithsonian Archives who enthusiastically supported my admittedly quirky research. I also want to sincerely thank Walter Tschinkel, R. O. Lawton Distinguished Professor Emeritus of Biological Science at Florida State University, who very generously provided funds to support my research. Finally, I put the finishing touches on this book while serving as a Postdoctoral Fellow at the Consortium for History of Science, Technology, and Medicine, and I am indebted to Babak Ashrafi, Amanda Casper, Roberto Chauca Tapia, Heidi Hausse, Phillip Honenberger, Julia Mansfield, Lynnette Regouby, Sheila O'Shaughnessy, and Aaron Slater for their support, as well as one exceedingly generous person who funded my fellowship anonymously.

I had the privilege of personally meeting with many different people while researching this project. For example, I discussed feral animals with a variety of people at several different academic conferences, and I am grateful to all of them. I would especially like to thank my various co-panelists, including Diana Ahmad, Joanna Dean, Ann Norton Greene, Jason Kauffman, Joshua Kercsmar, Scott Mittenberger, Brett Mizelle, Edmund Russell, Samiparna Samanta, Sandra Swart, and Sam White. In addition, many other people graciously allowed me into their respective homes, offices, or laboratories, and patiently answered all of my questions. I want to express my sincere appreciation to I. Lehr Brisbin at the Savannah River Ecology Laboratory, Jack Mayer at the Savannah River National Laboratory, Melinda Zeder and Torben Rick at the National Museum of Natural History, Elizabeth Reitz at the University of Georgia, Glen Doran and Rochelle Marrinan at Florida State University, Skip Snow at Everglades National Park, Kate Christen at the National Zoological Park, Phil Sponenberg at Virginia Tech, Brooks Miles Barnes at Eastern Shore Public Library, Xiomar Mordcovich at Miami-Dade Animal Services, Elizabeth DuBose at the Ossabaw Island Foundation, as well as Bernard Unti, John Hadidian, Andrew Rowan, and Wayne Pacelle at the Humane Society of the United States. I also need to thank all of the people whom I never chanced to meet, but who graciously answered my questions on the phone and over email. Many thanks to Denise Bowden at the Chincoteague Fire Department, Peter Savolainen at the Royal Institute of Technology (Stockholm), Doug M. Hoffman at Cumberland Island National Seashore, Karen McCalpin and Wesley Stallings at the Corolla Wild Horse Fund, Sue Stuska at Cape Lookout National Seashore, Luis Salgado at Miami-Dade Animal Services, and Jeffrey P. Blick at Georgia College and State University. In addition to all these folks, plenty of people answered my questions and I simply never took down their names. Many thanks to the helpful staff members at all the various archives, libraries, and coffee shops that I have visited over the past few years.

This book began as my dissertation, and I had the good fortune to meet some truly awesome people while living under the Live Oaks and Spanish moss in Tallahassee. First, I want to thank Cindy Ermus. She has listened to me ramble on and on about dogs, horses, and pigs for years, and yet she has somehow never lost her charm, her wit, or her mind. I can never thank her enough. I also want to thank Justin Black, Madisen Rard, Scott Shubitz, Tiffany Hensley, Hendry Miller, Chris Wilhelm, Jonathan Shepard, Weston Nunn, Josh Meeks, Bryan Banks, Thomas Lahr, and Mike Bartholomew for our many fruitful conversations over the years.

And since it took me more than thirty years to write my first book, and I don't know when I'll write another, I want to give a shout-out to my brothers and best friends, Bays, Josh, and Jacob, their wives, Ranessa, Christina, and Kim, and their kids, Addison, Ainsley, Lera, Miles, Silas, Fisher, and Lucas. I love them all and they all deserve to see their names in print. Last but not least, I want to thank my parents, William Bentley Gibson and Martha Sue Wolfe Gibson, who gave me life and then taught me that the world is beautiful. They are both poets at heart. I thank them for everything they have ever done for me, and I dedicate this book to them, with love.

# 1

## THE TROUBLE WITH FERALITY
### DOMESTICATION AS COEVOLUTION AND THE NATURE OF BROKEN SYMBIOSES

*Things fall apart, the centre cannot hold ...*
~ William Butler Yeats

Few enterprises have influenced the course of human civilization more than the domestication of animals. Our ancient bond with domestic animals predates many of the most important milestones in human history, including the development of agriculture, the establishment of cities, and the invention of writing. With animals by our sides, we became better clothed, better fed, and better warriors. We became stronger, faster, and more mobile. Domestic animals have accompanied humans across the globe since Paleolithic days, and they have helped us dominate every biome that we have encountered along the way. Meanwhile, the partnership has proven no less transformative for the animals involved. They are now distributed around the world in cosmopolitan fashion, and they have terraformed the planet almost as much as we have. From an evolutionary perspective, the bond has proven exceedingly advantageous, and domestic animals now vastly outnumber their closest wild brethren.

To understand how domestic animals have transformed human populations, and vice versa, one must recognize three salient facts. First, the domestication of animals was not something that happened a long time ago. On the contrary, it happens every day. You see, more than anything else, domestication is a *relationship*. In this case, it is a relationship among billions of partners from multiple species lasting thousands of years, but it is a relationship just the same. And, like any relationship, it requires sustained effort if it is going to flourish. Nothing is guaranteed. The ancient covenant must be forged anew every generation – or not. This leads to our second fact: things fall apart. Not always, but sometimes, and when they do, it is not uncommon for the parties involved to go their separate ways. If any erstwhile domestic animals survive their breakup with humanity and successfully establish residency in the "wild," free from direct anthropogenic selection, those animals and their progeny are technically neither wild nor domestic, and are instead relabeled *feral*. Thus our third and final fact: If we really want to understand our complicated relationship with domestic animals, then we cannot simply ignore the ones who went feral.

Like their closest cousins, feral animals were once joined with humanity in the coevolutionary embrace known as domestication, but something went wrong and the bond disappeared. This fact renders them unique among life on Earth, for they alone have known both our warm embrace *and* our cold shoulder. Fair or not, they are defined by what turned out to be a fleeting tryst. This perspective has left them with a fascinating, if not always flattering, perspective on human history. For thousands of years, they have stalked the periphery of human habitats, haunting flesh-and-bone reminders of a broken symbiosis, bearing silent witness to our worse and better natures. This is more than just literary hyperbole, by the way. It is also a scientific fact. Domestication affects an animal's physical and psychical constitution in profound ways, and humanity's fingerprints remain inscribed on the animal's genome long after the coevolutionary partnership breaks down. As a result, feral animals are subject to unique selective pressures quite distinct from those shaping their closest domestic and wild conspecifics.[1]

This is by no means the first work to examine ferality. As a matter of fact, biologists have recognized feral animals as discrete objects of scientific analysis for hundreds of years. When the French naturalist Comte de Buffon belittled North America's supposedly degenerate fauna in the late eighteenth century, he cited the continent's feral animals among his evidence.[2] In response to this sleight, Thomas Jefferson, a proud son of the New World, protested that America's feral animals were no more degenerate than those in Europe.[3] A century later, British naturalist Charles Darwin famously cited domestic animals as evidence in favor of natural selection, but what is less well known is that he also cited feral animals.[4] So too did his countryman, the co-discoverer of natural selection, Alfred Russell Wallace, when he independently articulated the exact same principle.[5] As concern for global biodiversity intensified during the twentieth century, many biologists quit celebrating feral animals as exemplars of evolution and began disparaging them as "invasive" pests who ought to be destroyed.[6] Most biologists continue to advocate for (or consent to) the annihilation of feral populations around the globe, though some scientists now champion the protection of some feral populations for varying reasons.

Despite all the attention that scientists have devoted to feral animals over the years, they have still not resolved what, precisely, qualifies as a feral animal. Most apply the feral label to any animal who was once domesticated, or had domesticated ancestors, but who now lives in the "wild." Note that this definition assigns feral status to both the initial founder strays and all of their progeny thereafter.[7] According to this definition, feralization begins in that moment when two previously engaged actors start to drift apart and the bonds between them start to disappear. Some researchers have described the feralization process as "domestication in reverse," but that is not entirely accurate. After all, ferality is really more of a *post*-domestication state.[8] Others think we ought to reserve the feral designation for animals who live and reproduce completely independent of humans, and thereby distinguish true feral animals from commensal "pariahs" who live in human-built environments and who subsist on human scraps.[9] Some complain that the feral label fails to convey the profound behavioral changes that attend feralization, and therefore use the "wild" adjective when describing once-domesticated populations.[10] Still others express concern that, unlike "wild," which has ennobling connotations, "feral" has negative connotations, relegating feral animals to the "low status of ecological pests"[11]

Biologists are not the only professionals who have shown an increased interest in feral animals over the years. Historians have also produced an impressive body of scholarship analyzing the unique role that feral populations have played in human history. Alfred Crosby has discussed their influence in several works, revealing that feral animals helped Europeans colonize other continents beginning in the fifteenth century.[12] Elinor Melville has shown that feral sheep transformed landscapes and societies in sixteenth-century Mexico.[13] Virginia DeJohn Anderson's research has demonstrated that feral animals helped English colonists establish and expand their presence in eastern North America throughout the seventeenth century.[14] Meanwhile, Harriet Ritvo's work on the Chillingham cattle of northern England has proven that feral populations often served as vehicles for human desires in the eighteenth and nineteenth centuries.[15] These classic texts helped establish feral animals as viable subjects for historical inquiry, and a growing number of writers have now begun to probe the complicated nature of ferality.[16]

This book attempts something slightly different. The narrative traces humanity's ancient relationship with domestic animals, the extent to which that relationship has experienced feralization over the years, and the evolutionary consequences for all parties involved. Its emphasis on evolution places the project squarely within the burgeoning field of study known as "evolutionary history," which examines how humans and other species have shaped each other's gene pools over the years. This method of analysis acknowledges that humans have shaped the evolutionary fate of other species (intentionally or otherwise) for millennia, and that this human-induced evolution has, in turn, transformed human history. In other words, human and nonhuman populations have, in a very real sense, coevolved with one another. The coevolutionary hypothesis has proved especially popular among those who study domestication. Michael Pollan famously employed this method to great effect when studying domestic plants in *Botany of Desire*, though others have since applied the same method to domestic animals.[17] Edmund Russell, who was the first to truly define evolutionary history as a distinct field of inquiry, has examined the evolutionary history of animals in several publications, and numerous others have since followed suit.[18] These works are important contributions to historical scholarship, but none of them focus on domestication's ephemerality, and none of them satisfactorily answer the still-pressing questions: What happens when our braided coevolutionary trajectories start to unravel? What happens to the animals we no longer need?

To assess the evolutionary effects of domestication and feralization on a given population, the narrative tracks three different variables over time: biogeographical distribution, genetic composition, and behavioral engagement with humans. These variables reveal that the distinction between domesticity and ferality is never clear-cut, and that humanity's innumerable interactions with other animals fall somewhere on a spectrum between total engagement and total indifference. In the case of biogeographic distribution, animals who live in enclosed spaces like crates, pastures, or apartments would fall closer to the domesticated end of the spectrum than animals who walk wherever they please, without any sort of human-imposed restrictions. The same holds true for genetic composition. Some populations are subject to intense anthropogenic manipulation in the name of sustenance, recreation, or vanity, while others reproduce of their own accord, without any input from people. Finally, tracking each population's behavioral engagement with humanity is essential to demarcating domestic and feral animals, but drawing distinctions is not always easy. Some animals have never willfully engaged humans, yet they owe their existence to anthropogenic intervention. Others are carefully chosen and then deliberately abandoned, all so they can better emulate their pre-domesticated ancestors.

*The trouble with ferality* 5

In cases like these, it is not at all clear whether the animals in question are wild, domestic, or feral.

Studying ferality also provides one with a fresh perspective on the long-running wilderness debate. As environmental historians of North America are well aware, the idea of wilderness has meant a great many things to a great many people over the years. The earliest European visitors considered the entire continent an unsettled wilderness, despite the fact that millions of people already lived there. They viewed the continent as something to be conquered and dominated, and they proceeded accordingly. Then, at precisely the same time that the "frontier" closed, saving the last remnants of wilderness assumed paramount importance among the nation's most urbane individuals. Consequently, when Roderick Nash examined the *idea* of wilderness in 1967, he placed nature and culture on opposite ends of a spectrum.[19] A generation later, William Cronon noted that viewing wilderness as something apart from humans leaves no room for humans, and therefore undermines our efforts to exist *with* nature.[20] Most of the scholarly discussion about wilderness since then has highlighted the "hybrid" nature of wilderness, which is at once natural and constructed.[21] This project demonstrates that the same false distinctions between "wild" and "artificial" apply to animals no less than landscapes. Determining whether a given population qualifies as wild, domestic, or feral requires that one tease apart the animals' respective evolutionary histories, and this, in turn, requires familiarity with the historical record.

In an effort to limit what would otherwise prove an impossible scope, the narrative limits its analysis of domestic animals to three of history's most conspicuous species: dogs,[22] pigs,[23] and horses.[24] Each of these animals entered domestication at different times, in different places, and under drastically different circumstances. Chapter 2 introduces each of these animals, and then explains how they first became acquainted with humans. Though people and animals independently entered domestication at many different times and in many different places (including the Americas, Africa, and Oceania), dogs, pigs, and horses all entered domestication in prehistoric Eurasia.[25] More precise details are not always forthcoming, but combing the archaeological record and the genetic data sheds considerable light on the origins of domestication. These analyses demonstrate that the reproductive boundaries distinguishing wild, domestic, and feral populations remained relatively ill-defined for several thousand years.[26]

This emphasis on the origins of domestication should also improve our understanding of humanity's larger relationship with the rest of nature. After all, scientists and scholars agree that humans are the primary force shaping the evolution of life on Earth, and that we now live in a new, human-dominated geological epoch known as the Anthropocene. Even so, it is not at all clear when this so-called epoch actually began. Many believe that it began in the eighteenth century, when the Industrial Revolution started producing major atmospheric changes. Others insist that humanity's impact was not really felt on a global scale until the mid-twentieth century, but that human activities have greatly accelerated since the end of World War II. Still others have pushed the needle in the opposite direction, noting that humans have engineered their ecosystems for millennia, and that the origin of agriculture thus denotes the beginning of the Anthropocene. This project agrees that domestication provides the most explicit demarcation of the human-controlled world, but it reminds readers that the domestication of animals preceded the domestication of plants by thousands of years.[27]

Subsequent chapters recount how humans have interacted with each of these three species in one specific place: the American South.[28] Since this region is notoriously difficult to define, a few extra words are perhaps in order. After all, any study that purports to examine the American South probably ought to examine the entire southern half of the modern-day United States (including, presumably, Hawaii), but they never do. Historians have suggested that we study the southern half of the continental United States as a single region, a Sun Belt, but relatively few scholars have answered the call.[29] When most people talk about the South, they are talking about the southeastern quadrant of North America. Pressed for specifics, they would probably equate the South with the eleven states that briefly formed a Confederacy in the early 1860s.[30] Sure enough, most environmental histories of "the South" focus on the southeastern quadrant.[31]

Tradition notwithstanding, this book draws a still tighter geographic focus. The narrative trains its attention on that part of North America that lies east of the Appalachian Mountains and south of the Potomac River. This region is composed of modern-day Virginia, North Carolina, South Carolina, Georgia, and Florida (see Figure 1). There are legitimate reasons for treating the region as a distinct ecological theater. For example, the five states share more than a thousand miles of coastline along the Atlantic Ocean, and this has affected the region's culture, ecology, and identity in profound ways. Meanwhile, the Appalachian Mountains

Figure 1 Map of southeastern North America
This book focuses on southeastern North America. More specifically, it focuses on the area east of the Appalachian Mountains and south of the Potomac River. This area is composed of five modern-day states (Virginia, North Carolina, South Carolina, Georgia, and Florida) and three physiographic zones (ancient highlands, piedmont plateau, and coastal lowlands).
**Source** – *The National Atlas of the United States of America* (Washington, D.C.: United States Department of the Interior, 1970), 56–58. Map Collection, Perry-Castañeda Library, University of Texas at Austin.

have influenced settlement patterns, agricultural practices, and trade networks in southeastern North America for thousands of years. As a result, communities to the west of the mountains have developed in subtly different ways from those to the east.[32] In similar fashion, communities in the Gulf of Mexico watershed have developed differently from those in the Atlantic Ocean watershed.[33] For these reasons, the narrative focuses on the region east of the Appalachians and south of the Potomac, though it should be noted that this region is by no means homogenous. It contains three distinct physiographic zones (ancient highlands,[34] piedmont plateau,[35] and coastal lowlands[36]), and while most of the region is temperate and experiences four distinct seasons, the southern half of Florida is subtropical and experiences just two seasons (dry and wet).[37] Finally, readers should note that while the narrative refers to this region as "the South" for the sake of brevity and convenience, it readily acknowledges that there are, in fact, many Souths.[38]

Because none of the book's four main characters are native to the American South, Chapter 3 explains how they all came to populate the region. People and dogs first arrived in the region more than 11,000 years ago, and they were joined by pigs and horses (and still more people and dogs) in the early sixteenth century, but it was not until the English intensified colonization efforts in Virginia during the early seventeenth century that the number of domestic and feral animals began to skyrocket. Colonists famously encouraged their livestock to roam free, showing virtually no regard for the Native Americans who already lived there, or their quite legitimate claim to first dibs on everything. Meanwhile, dogs were deeply engaged with people of every stripe during the colonial period, but neither their range nor their reproduction was closely monitored. When additional settlers established additional colonies in the Carolinas, Georgia, and Florida, the exact same patterns were repeated, with subtle regional variations. By the late eighteenth century, feral animals drastically outnumbered domestic animals in certain places, a fact that meant different things for the region's rapidly changing collection of Native American, European, and African peoples.[39]

Studying feral animals during the colonial period also allows one to engage the large body of literature on the so-called southern frontier.[40] More than eighty years after his death, historian Frederick Jackson Turner remains the looming figure in this historiography. According to Turner, the southern frontier was defined by the same settlement patterns that defined the frontier elsewhere. Namely, the trading frontier gave way to the livestock frontier, which gave way to the agriculture frontier, which gave way to the industrial frontier.[41] Historians now dismiss Turner's interpretation as too ethnocentric, but his thesis remains a lightning rod all these years later. Allan Greer suggests that we redefine Turner's frontier as "the zone of conflict between the indigenous commons and the colonial (outer) commons."[42] Others suggest that we relabel the frontier something else entirely, like "the commons," "the backcountry," "the interior," "the shatter zone," and "the borderlands."[43] Meanwhile, historian Mart Stewart reminds us that the frontier means something entirely different in the South, where wilderness has always been more proximate, where nature has always been "*inhabited.*"[44]

Chapter 4 examines the peculiar customs and institutions that prevailed in the South from the early national period to the end of the Civil War (roughly 1783 to 1865). More densely settled parts of the United States had long since abandoned open-range husbandry, but the practice continued without abatement in the South, with different implications for different species. As John Majewski and Viken Tchakerin recently demonstrated, the large tracts of "unimproved" land that covered the region during the nineteenth century were not untouched wilderness that had never been utilized. Instead, most of that land had already been intensively cultivated at some point in the recent past and then allowed to lie fallow for decades. This method of shifting cultivation not only created the "demographic equivalent of a permanent frontier," but also provided feral animals with habitats in which to live.[45] What is more, all people in the South engaged with feral populations during this era, no matter their ancestry, but their interactions with the animals revealed unequal power structures among humans.[46]

Significantly, the reproductive distinctions separating domestic and feral animals started to grow more pronounced during this period, and, as a result, populations started to cleave. Invariably, some members of each species were targeted for increased anthropogenic manipulation, and this meant new selective parameters. A growing number of horses, pigs, and dogs were specifically cultivated to serve a variety of technological, agricultural, and martial purposes. Despite these initiatives, however, many animals remained free-ranging and genetically undifferentiated during the antebellum period. Once again, this meant different things for different species, each of which boasted unique ecological footprints and unique rates of fecundity. Feral populations were alternately subdued, ignored, or resented. The populations of all animals, both domestic and feral, decreased dramatically in the early

1860s. Census figures show that the region's supply of pigs and horses diminished by a third during the Civil War.[47]

Chapter 5 examines the period from the end of the Civil War to the end of World War II (1865–1945), a period characterized by increased industrialization, mechanization, and urbanization. These processes developed more slowly in the South than other parts of the nation, but they developed just the same. The growing number of people who lived in the region (most of whom were descended from Europeans, Africans, or both) became increasingly entrenched in national markets and gene flows, with quite different implications for the region's animal populations. The region's wealthiest planters tried to close the open range on numerous occasions following the wake of the Civil War, but they were consistently thwarted by an unlikely coalition of "poor whites" and "poor blacks," who continued to range their livestock on the commons. This fact is significant. Scholars sometimes assume that race has always trumped class in the South, but the debate over fencing livestock (especially feral pigs) shows that people of different ancestries often engaged animals in similar ways. In due course, however, human population densities continued to increase and, slowly but surely, the open range collapsed in piecemeal fashion.[48]

Meanwhile, industrialization drastically increased demand for horses, and, as a result, the number of equids living in the region continued to grow throughout the late nineteenth century. Despite this growth, the southeastern states had still not recovered their prewar complement of horses by 1900. Wartime devastation was partly to blame, but the bigger culprit was the region's preference for horse/donkey hybrids known as mules. In the early twentieth century, revolutionary advances in technology and transportation devastated the region's domestic equids. These developments meant different things for their feral cousins. Some were prized more than ever. Others were gunned down and destroyed. Finally, many of the wealthiest people in the South began spending exorbitant sums of money cultivating ever more refined domestic dogs, even as the region's rapidly expanding cities waged an ongoing extermination campaign against their cultivars' admixed, free-ranging cousins.[49]

Having reviewed the historical conditions that created and later sustained feral populations over the past several centuries, Chapter 6 examines how dogs, pigs, and horses are faring in the modern South (1945 to the present). The reproductive boundaries distinguishing their domestic and feral representatives are more pronounced, and thus more evolutionary significant, than ever before. People in the South spend millions of dollars housing, feeding, and sometimes even clothing their dogs, yet they also slaughter their pets' feral cousins on a daily basis. Domestic horses have grown more scarce, a fact that has, somewhat counter-intuitively, imbued their closest feral cousins with greater value. Meanwhile, domestic and feral populations of pigs have both exploded of late, but for drastically different reasons. This chapter demonstrates that these gene pools vary in response to humanity's actions, conscious or otherwise.

A brief Epilogue asks whether there is any room for feral populations in the current human-dominated epoch of world history known as the Anthropocene. After all, feral animals are defined by their distance from humans, and yet one of the central precepts of the Anthropocene is that no part of the planet, and thus no animal population, is free from anthropogenic influence. The book closes by reviewing the evolutionary history of each feral population, forecasting their respective evolutionary trajectories, and dispassionately assessing whether domestic or feral animals have the more preferable destiny.

Before proceeding to the narrative portion of the text, we must address a few technicalities. Strange though it may sound, scholars do not really have a good name for the people who were already living in the Americas prior to the arrival of Europeans. They once referred to these people as "Indians," though the preferred nomenclature these days is "Native American." This label has its own obvious faults (according to scientists, no human is truly *native* to the Americas), but it recognizes the legitimate primacy of the first Americans and their descendants. This project will use the "Native American" label, but it will also bow to popular convention and use the "Indian" label. Within this study, the labels are treated interchangeably.[50] Another technicality regards spelling in Chapter 3. Although colonial settlers generally spoke and wrote in English, their spelling was wildly inconsistent. I have therefore taken the liberty of editing their spelling for modern readers. While we are on the topic of colonialism, readers will note that "English" settlers become "British" settlers about halfway through the chapter, a fact that reflects the offstage unification of Great Britain in 1707.

Finally, a few clarifying words about the definition of "feral" are also in order. The reader may have noticed that there are two acceptable definitions of the word (see page viii). The first definition applies to domestic animals who have taken leave of humanity's immediate dominion and established residency in the wild. The second definition is more of a behavioral descriptor (feral = "wild; savage; ferocious"). This project documents the anthropogenic factors that have influenced the genetic composition and biogeographic distribution of animal populations over time, and, as a result, the narrative is not terribly concerned with the second definition, except inasmuch as it influences each group's evolutionary fortunes. Also, one last thing: I use the word "ferality" throughout this project to describe the state of being feral, but I confess that the word does not otherwise exist. According to *Webster's Dictionary*, the word "feral" is derived from the Latin word *ferus* (meaning "wild"), and one should therefore use "ferity" when referring to the state of being feral. I nevertheless prefer to use "ferality" for two reasons. First, it is phonetically more intuitive. But the second and more important reason is that ferality more readily calls to mind the Latin word *feralis* ("funereal; of or relating to the dead"). Given the unknown fates that often await the animals we no longer need, I believe this connotation is altogether more fitting.

## Chapter Summary

For thousands of years, people have divided the world's vertebrate fauna into two distinct categories, wild and domestic, but some animals cannot be shoehorned into this strict dichotomy. Feral animals (erstwhile domestic animals who now live in the wild) represent a third category of life. They are neither domestic nor wild, but have instead created a post-domestication niche. Like domestication, ferality is a cultural category that denotes an organism's proximity to (or distance from) humans, and, like domestication, ferality has profound evolutionary consequences. To demonstrate as much, this book examines the evolutionary histories of three different species (dogs, pigs, and horses) in one particular region (southeastern North America), especially the extent to which they have experienced ferality over the years. The South is not the only region that has hosted feral animals, but its warm climate and peculiar settlement patterns have sustained large feral populations in one form or another for the past several centuries. The species are likewise deliberately chosen, not only because of their conspicuous roles in the region's history, but also because they have had such radically different experiences with both domestication and ferality. As we shall see, the boundaries distinguishing various populations were sometimes so fine as to seem negligible, and yet those distinctions often spelled the difference between life and death, between existence and extinction.

# 2

## MAKING AND BREAKING ACQUAINTANCES

### THE ORIGINS OF WILDNESS, DOMESTICATION AND FERALITY IN PREHISTORIC EURASIA

*Friendship is born at that moment when one person says to another, 'What! You too? I thought I was the only one!'*

~ C. S. Lewis

According to the modern scientific consensus, anatomically modern humans (*Homo sapiens*) first evolved in eastern Africa about a quarter-million years ago. The oldest indisputably human fossils were found near the Omo River in southern Ethiopia, and date back more than 200,000 years. Similar fossils have turned up in the nearby Awash Valley, also in Ethiopia, that date back approximately 160,000 years, thereby lending support to the hypothesis that *Homo sapiens* first evolved in the Horn of Africa.[1] Attempts to trace humanity's subsequent dispersal throughout the rest of the continent soon run into difficulty, however. Thanks to an incomplete fossil record and patchy climatic data, scientists are unsure whether the *Homo sapiens* who first evolved in Ethiopia overwhelmed and displaced their hominid brethren elsewhere in Africa, or whether there was instead significant admixture.[2] In either case, members of our own species were the only hominids left in Africa by 140,000 years ago.[3] This is a matter of no small significance. However different we may be, all humans alive on Earth today trace our ancestry back to this original African wellspring.

Not long thereafter, some humans began venturing beyond their African homeland. They first entered southern Arabia by crossing the Red Sea at the Bab-el-Mandeb Strait more than 130,000 years ago, and later filtered into modern-day Israel by crossing the Sinai Peninsula some 115,000 years ago.[4] Most researchers agree that these early pioneers and their descendants contracted back toward Africa, or died out entirely.[5] Subsequent dispersals proved more successful. Around 60,000 years ago, scattered groups of people left Africa, and these pulses effectively seeded the rest of the world with humans. Some people trekked east toward the

Indian subcontinent, reaching Australia around 50,000 years ago, while others slowly diffused across the vast Eurasian landmass.[6]

The evidence suggests that people reached Europe around 45,000 years ago, and that they reached northeastern Siberia by 30,000 years ago.[7] Because the Earth was much cooler at the time, and the planet's eustatic water table was therefore significantly lower, Siberia was connected to Alaska as part of a single land bridge known as Beringia. The first visitors to Beringia might have otherwise strolled through Alaska and discovered the rest of the Americas, but the entirety of Canada lay beneath the sprawling Laurentide ice sheet. Enormous glacial walls prevented the region's earliest hunter-gatherers from migrating into the Americas (or even knowing that the continents existed) for thousands of years thereafter. As a result, the land "bridge" was actually more of a land "cul-de-sac."[8]

As humans swept across Eurasia, they encountered two other hominid species, Neanderthals and Denisovans. These hominids were not quite like you and me, but they were not so different either. They controlled fire, coordinated hunting trips, and doodled on nearby objects.[9] As one might expect, the fate of these early hominids remains somewhat controversial. Scientists once believed that *Homo sapiens* outcompeted and replaced their cousins, but most now allow for some level of interbreeding among the three species. Whether killed off or subsumed, humanity's closest sister species were reduced to relict populations by 40,000 years ago, and had disappeared entirely by 27,000 years ago.[10]

It was also around this time, some contend, that our species began to manifest greater regional variation. Over the past few years, numerous scientists have attempted to identify when, precisely, the planet's various "races" first took shape. So far, these efforts have proven inconsistent and inconclusive. For example, they have produced a wide range of dates for the purported racial divergence of Asians and Europeans, including estimates of 14,000–20,000 years ago,[11] 17,000–43,000 years ago,[12] 23,000 years ago,[13] 25,000 years ago,[14] 31,000–40,000 years ago,[15] 36,200 years ago,[16] 39,000 years ago,[17] 40,000 years ago,[18] 44,000 years ago,[19] and 40,000–80,000 years ago.[20]

As these rather imprecise dates imply, attempts to calculate the purported evolutionary divergence of human races are inherently problematic and perhaps impossible. After all, most scientists and scholars agree that human races are not even real. More precisely, they insist that race is a social construct, a set of utterly contingent cultural assumptions that we inherit through our experiences but which lack any ontological justification. A few insist that biological races are absolutely real and evolutionarily significant, but their models generally fail to account for the fact that all humans were still exclusively nomadic throughout the long Pleistocene.[21] This mode of life allowed disparate peoples across Eurasia to interbreed with one another and recombine their respective genes. It also allowed disparate peoples to migrate from one continent to another, including migrations back into (and out of) Africa.[22] As one group of geneticists recently explained, genetic admixture has been an "almost universal force shaping populations."[23] Therefore, despite historical attempts to parse and divide humans (a phenomenon that would later reach acute levels in southeastern North America), the various "races" of humanity in fact represent little more than transient genetic eddies subject to random biogeographical variation over time.

Meanwhile, it was at some point during humanity's headlong diffusion across the Eurasian continent between 40,000 and 15,000 years ago that some members of our species first entered into domestication with another species. Particulars surrounding the initial timing, location, method, and meaning of domestication remain the subject of sharp debate, but everybody can agree on one thing: dogs were definitely first. We should not be terribly surprised that our best friend was also our first friend. Humans commune more closely with dogs than any other species, and this has proved true since time immemorial. Our relationship with these sociable creatures not only predates our species' collective memory, but also predates our expansion into the Americas. Simply put, dogs have been part of the human experience for a very, *very* long time.

It has not always been clear which canid species served as the progenitor to dogs. In the eighteenth century, the oft-maligned Jean-Baptiste Lamarck suggested that it might have been wolves, while his more celebrated successor, Charles Darwin, later wrote that it might have been jackals.[24] Debate persisted throughout the twentieth century, but increasingly refined genetic analyses have finally established that the dog's closest non-domesticated cousin is the gray wolf (*Canis lupus*). Note the deliberate construction of the previous sentence. Dogs are not "descended" from wolves. That kind of language misrepresents the evolutionary process. It inaccurately suggests that wolves have stopped evolving and that they are somehow arrested in time. Extant wolves can hardly be the ancestors of extant dogs since both parties are, after all, extant. It is more accurate and thus more meaningful to acknowledge that wolves and dogs are cousins who share a relatively recent common ancestor. In fact, the two species are so closely related that many now categorize dogs as *Canis lupus familiaris*.[25]

Trying to determine *when*, *where*, *how*, and *why* wolves and dogs began to diverge from one another has always proven more difficult. For several decades, scientists generally agreed that canid domestication began around 12,000 to 15,000 years ago.[26] That assumption was turned on its head in 1997, when a team of geneticists published results indicating that wolves and dogs may have begun to diverge from one another as early as 135,000 years ago.[27] They intimated that canid domestication might have therefore begun more than a hundred millennia earlier than previously thought. Such an ancient origin for dogs would have many implications. For example, if dogs have been around for more than 100,000 years, then there is a very real possibility that Neanderthals and Denisovans may have had some hand in their domestication. This would radically transform our understanding of early human history. Domestication is one of the reliable hallmarks that paleoanthropologists use to distinguish behaviorally modern humans. Assigning that distinction to now-extinct hominids would necessarily challenge our idea of what it means to be "human."

Several recent studies have helped shed light on the matter. Geneticists recently sequenced a variety of canids from around the world, studied the genetic variation among them, and concluded that dogs were first domesticated around 32,000 years ago.[28] Other genetic studies have provided comparable estimates.[29] Meanwhile, the archaeological evidence suggests a similar timeline. Over the last few years, archaeologists have identified numerous dogs in the fossil record that push the purported date of domestication back by thousands of years. One team has identified several "proto-dogs" scattered throughout central Europe that are more than 27,000 years old, while another team has recovered an "incipient dog" from southern Siberia that dates back more than 33,000 years.[30]

Just as the initial timing of dog domestication has inspired debate, so too has its original location. Many insist that dogs first diverged from wolves in that part of China lying south of the Yangtze River, but that opinion is not universal. Several other high-profile studies have claimed that dogs entered domestication in Europe, Mongolia, and the Middle East.[31] Since wolves were coterminous with people in all of these places during prehistoric times, it is not unreasonable to think that domestication commenced in multiple places multiple times. Meanwhile, further testing will be necessary to show whether the oldest dog remains in ancient Africa signal an independent domestication event from local wolves, or whether they instead suggest that foreign travelers first introduced dogs to Africa at some point in the continent's ancient history.[32]

Ultimately, the *when* and the *where* of initial domestication are secondary to the more interesting questions, *how* and *why*. To address these questions, one must first account for what might have possibly drawn these two species together. It is significant, for example, that whether domestication commenced 30,000 years ago or 15,000 years ago, all humans everywhere were nomadic hunter-gatherers. Throughout Africa, Eurasia, and Australia, people followed game. As a result, many researchers believe that something about the hunting lifestyle must have thrust these two species together. Because many of the earliest dog specimens were found among mammoth-hunting peoples, for example, some experts insist that canids might have helped in hauling meat and equipment. These scholars recognize that it would require a lot of meat to sustain the earliest packs of domesticated dogs, but they insist that mammoths would have provided more than enough.[33] Others think it had less to do with the dogs' hauling capacity and more to do with their own impressive hunting prowess. These researchers emphasize the fact that Pleistocene-era humans and wolves generally pursued the same type of game and employed similar hunting tactics, and might have found it more useful to coordinate with one another rather than compete.[34]

Others promote a different origin story. Ashley Montagu was among the first to suggest that early dogs may have been scavengers who pilfered scraps from the periphery of human habitats, but he was not the last, and researchers from a variety of disciplines continue to produce evidence in support of this hypothesis more than seventy years later.[35] Behavioral ecologists Raymond Coppinger and Lorna Coppinger endorse a scenario in which dogs began their association with people as "dump invaders," while historian Edmund Russell hypothesizes that people might have tolerated scavengers who reduced the stench of rotting meat around camp. Killing the most "cantankerous" wolves would have eventually produced tamer populations of wolves, he explains, which would in turn help facilitate domestication.[36] Finally, archaeologist Melinda Zeder writes that selecting for less-aggressive animals could also help explain why skeletons of ancient dogs manifest greater juvenilization than wild conspecifics.[37]

These views all explain *how* domestication may have first occurred, but not *why*. The conventional way of thinking about domestication presumes, as a matter of course, that humans were responsible for initiating the relationship. *Domesticate* is a transitive verb, after all. It was people who acted upon passive animals. As Pierre Ducos once explained, domestication "exists because humans (and not the animal) wished it."[38]

A growing number of researchers have begun to reject this mode of thinking, however. They insist that domestication is best understood as a coevolutionary process, and that there is no reason to believe humans were solely responsible for getting things started.[39] This perspective accords well with new research that proves dogs are more finely attuned to human communication cues than any other species, and that selection may have favored "human-like social cognition" during the earliest stage of domestication.[40]

If it is true that a greater tendency toward inter-specific social engagement distinguished the earliest dogs from their lupine contemporaries, one still has to explain the apparent *morphological* changes that accompanied domestication. After all, skeletal differences are one of the primary ways that archaeologists distinguish fossil dogs from fossil wolves.[41] A decades-long experiment in southern Siberia may provide the answer. Back in the 1950s, Russian geneticist Dmitry Belyaev began breeding silver foxes, a melanistic form of the common, wild (never-domesticated) red fox. He and his team bred the animals based on a single criterion: their tameness around humans. Those foxes who showed a greater willingness to engage with humans were matched with similarly inclined canids, and the same selective parameters were applied to subsequent generations as well. It was not long before Belyaev noticed something about the foxes' behavior. Each successive generation was more placid than the previous one. Furthermore, their behavior was starting to look eerily familiar. By the fourth generation, some of the kits (young foxes) had begun to wag their tails when in the presence of people, and whimper in hopes of being petted. By the sixth generation, the most gregarious foxes were licking the researchers' faces.

The emergence of dog-like behavior is remarkable, but not altogether shocking. After all, the researchers were selecting a specific behavioral trait. Little surprise if other behaviors were somehow correlated. What is far more remarkable is that the foxes eventually began to manifest physical changes as well. By the ninth generation, foxes were bounding around with floppy ears. Soon thereafter, they were being born with piebald coats, a phenotypic pattern common among many of the world's domestic creatures, but one that is absent in the wild. By the fifteenth generation, their fundamental structure had begun to change, as some foxes were born with between three and six fewer vertebratae.[42] In light of these suggestive correlations, many researchers now believe that the same suite of genes that controls an animal's fight-or-flight response must also control the animal's various morphological features as well.[43] Some have dubbed this unique morphological condition the "domestication syndrome."[44]

Note that domestication not only created domestic animals, but wild ones as well. As Roderick Nash explains, there was no meaningful distinction between humans and the rest of nature, "no dualism," until humans began engaging their local fauna in domestication. At that point, it became necessary to distinguish those animals who had entered domestication from those who had not. Wildness was originally identified in those animals who would not consent to domestication, and wilderness was, by definition, those places where wild, non-domesticated animals lived.[45] Domestication was not established in one fell swoop, though, and neither was wildness. Instead, the relationship was established over multiple generations, and that means that both people and proto-dogs had to somehow sustain their initial attraction. For years, researchers insisted that domestication required that the earliest dogs become genetically cloistered away from their still-wild cousins. On these grounds, some researchers have likened the origin of domestication to a speciation event.[46] Yet it is incorrect to assume that wolf puppies, once brought into the human camp, were automatically thereafter a new species. As numerous researchers have pointed out, dogs would still need some new genetic contributions to sustain the lineage. It therefore stands to reason that "wild" wolves would have continued to contribute their genes to "domestic" dog populations long after domestication first commenced. In fact, most researchers assume that dogs and wolves would have continued to exchange genes quite frequently during the initial stage of domestication, and that hybridization would have therefore been common.[47]

But that sword cuts both ways. If it is true that the reproductive boundaries distinguishing "wild" from "domestic" animals were really that permeable, it means that proto-dogs could have transgressed those boundaries just as easily as wolves, but in the opposite direction. Here now we see an important but underappreciated aspect of the domestication process. Namely, that the relationship, once initiated, was by no means guaranteed, and early "domestic" animals probably returned to the "wild" quite readily at first. Inasmuch as these animals removed themselves from anthropogenic selection, they became "feral." As a result, these animals challenge the notion that domestic animals represent new species. After all, domestication sometimes failed.[48] Most researchers now accept that there was significant genetic admixture among wild, domestic, and feral populations for the first several thousand years of animal domestication.[49]

Humans had already been engaged in a domesticated relationship with dogs for several thousand years when something remarkable started to happen. Around 12,000 years ago, for reasons that remain unclear, small bands of people who had only ever known a nomadic way of life began settling into semi-sedentary communities. Known as the Neolithic Revolution, this not-so-subtle shift in demography and behavior undoubtedly ranks among the most important events in human history. Putting down roots not only allowed people to build communities and develop trade networks, but also prompted them to invent government and coordinate specialized labor. And yet, despite these radical transformations, the most dramatic development was the extension of humanity's biological dominion. After all, it was not until people gave up the nomadic way of life and began settling into communities that they once again engaged other animals in coevolutionary partnerships.[50] And so the time has come to introduce the last two species, pigs (*Sus scrofa*) and horses (*Equus caballas*).

For years, understanding the events surrounding the initial domestication of pigs proved exceedingly difficult. For most other domestic species, scientists can limit their search for early specimens to the reconstructed range of each creature's progenitor population.[51] Not so for pigs. Wild suids first evolved in southeastern Asia millions of years ago, but these opportunistic creatures proved highly adaptable and they had spread throughout most of Eurasia by the time *Homo sapiens* joined them.[52] Archaeologists cite asymmetric demographic patterns among zooarchaeological remains as evidence that people had started manipulating local populations of free-ranging swine in the Near East (a region stretching from western Turkey to eastern Iran) as early as 10,500 years ago, and that domestication was therefore well under way. These pigs then spread west across Anatolia for the next several thousand years.[53]

It now appears certain that the earliest domestic pigs in Europe were brought there by farmers who began emigrating from modern-day Turkey around 6,500 years ago. Evidence shows that they migrated via two distinct routes, one band following the Danube valley, the other traveling along the northern coast of the Mediterranean.[54] These farmers brought a staggering array of plant and animal cultivars into a landscape that had heretofore known only hunter-gatherers. Plants included emmer and einkorn wheats, barley, peas, chick peas, bitter vetch, and lentil. In addition to plants, however, these people also migrated alongside herds of different animals, including cattle, sheep, goats and pigs. These early farmers diffused across Europe for the next several centuries. Evidence suggests that they had arrived in Portugal by 5,200 years ago, and northwest France by 5,000 years ago.[55] These farmers appear not to have "conquered" the native hunter-gatherers. Instead, evidence shows that both ways of life coexisted alongside one another for more than a thousand years. Genetic profiles from modern Europeans suggest significant interbreeding rather than antagonistic competition.[56]

It is worth repeating that when Near East farmers arrived in Europe, they did not find an area without suids of any kind. The reader will recall that wild (never-domesticated) boars were distributed throughout Europe. Things began to change around 5,500 years ago, when there was an abrupt shift in the genetic composition of Europe's pigs. Almost overnight, pigs of Near East ancestry disappeared from Europe, and pigs domesticated from European wild boars came to predominate. By 5,000 years ago, there were no more pigs of Near East ancestry left in Europe.[57] Remarkably, domestic pigs of European descent eventually spread into the Near East and supplanted the original population.[58] European farmers obviously influenced the European pigs' composition and distribution, but they did not otherwise develop any specific cultivars. That is because they practiced a unique method of husbandry known as "pannage," which encouraged pigs to scour the continent's expansive woodland floors in search of acorns and other mast. These pigs relied on their wits for survival and procreated beyond humanity's watchful eye. As a result, there was extensive hybridization among wild, domestic, and feral populations for thousands of years.[59]

Numerous studies have conclusively demonstrated that the people in the Near East and western Eurasia were not unique, and that people in India, China, the Philippines, and Africa proved equally willing to strike up a relationship with their local suids.[60] Comparing and contrasting these separate domestication events is enlightening, for doing so reveals that the process not only differed among species, but also differed among populations within species. For example, Chinese pigs and European pigs were subject to radically different selective pressures.[61] Unlike their European counterparts, Chinese pigs were far more likely to be confined in sties and fed at troughs at a very early stage in the domestication process, possibly as a result of higher population densities. Chinese farmers made special efforts to enclose domestic pigs and control their reproductive output, which necessarily influenced their genetic composition.[62]

Finally, the time has come to introduce horses, the last major livestock species to enter domestication. The horses' ancestors evolved on the broad plains of North America more than fifty million years ago, but some of these creatures trekked westward across the Bering Land Bridge around five million years ago. Those equids who remained behind in the Americas would eventually go extinct, while those who left achieved widespread distribution throughout Eurasia and Africa. Over the next several million years, these equids diversified into the extant categories we recognize today: zebras, asses, and horses.[63] These lattermost creatures were scattered across Eurasia by the time *Homo sapiens* joined them around 45,000 years ago.[64] Wild horses thrived during the frigid Pleistocene, when open conditions prevailed, but they struggled when the global climate warmed and Eurasia grew more arboreal. By 5,500 years ago, their range had been reduced to the central Eurasian steppes and the Iberian Peninsula.[65] They might have disappeared altogether had humans not intervened.

Identifying the first domesticated horses is not easy, especially since archaeologists cannot rely on traditional osteological evidence.[66] Unlike other livestock species, the skeleton of a domesticated horse is not significantly different from its coterminous wild brethren.[67] As a result, scientists and scholars must use other kinds of evidence. Citing metrical, dental, and organic-residue analyses, archaeologist Alan Outram has shown that the people of the Botai Culture in northern Kazakhstan possessed domesticated horses as early as 5,500 years ago.[68] The evidence suggests that they not only used these horses for milk and meat, but also locomotion. Riding domestic horses not only allowed these people to more effectively manage wild horses, but also increased efficiency when hunting elk, deer, antelope and aurochs.[69] Meanwhile, geneticists have shown that horses began to manifest much greater variation in coat colors around this time.[70] This kind of diversification is consistent with the origin of domestication and the subsequent relaxation of natural selection.

Partnering up with one another provided horses and humans with unprecedented advantages, and life on the steppes was completely transformed in relatively short order. The vast, empty plains stretching from eastern Hungary to western China, which had previously inhibited human migrations, suddenly burst alive as these strange equid–hominid symbionts began crisscrossing the high-altitude plains. Horses first accompanied steppe peoples into Europe around 5,000 years ago, filtering into the continent via the Danube watershed.[71] By 4,000 years ago, domestic horses had reached the British Isles. Meanwhile, on the other side of Eurasia, domestic horses helped establish the Silk Road, arriving in China around 4,000 years ago.[72] Traveling south from the Central Eurasian steppes, mounted nomads arrived in southern Iran around 4,000 years ago.[73] At first, they overwhelmed the larger but more sedentary civilizations in the Near East. In due course, however, the people of the Near East also adopted horses, who enabled rapid transport across the same desert wastes that had heretofore always inhibited sustained interaction. Over the years, as horses became more deeply integrated into societies throughout the Near East, they also began to change. These Arabian horses assumed lither, more gracile morphologies.[74] Humans no doubt encouraged this development, though Bergmann's Rule, which states that species in colder environments are generally larger than species in warmer regions, suggests that it was also the horses' evolutionary response to their environment.[75]

By 3,500 years ago, Arabian horses allowed Bedouin nomads to conquer all of North Africa, from Egypt to the Maghreb. Some traveled south, eventually filtering into western Africa.[76] Others crossed the Strait of Gibraltar into the Iberian Peninsula, where they encountered still more equids.[77] It is not entirely clear whether Iberians had already domesticated local horses prior to the arrival of domestic horses from abroad. In any event, there was significant admixture between the two populations. Wherever they trod, domestic horses recruited any relic bands of wild horses that they encountered, and thus continuously incorporated new genes from wild stock. As a result, domestication not only unfolded over long periods of time, but over large swaths of land as well.[78]

While researchers accept that domestic horses continued to incorporate genes from wild stock whenever they encountered them, it is not entirely clear whether the earliest domestic horses ever took leave of humanity's early dominion and contributed their feral genes to wild populations. There are no osteological signatures distinguishing feral horses from their coterminous wild and domestic cousins, and so researchers must consult other types of evidence. For example, archaeologist Marsha Levine has compiled a list of textual references to "wild horses" in Eurasia over the past 2,500 years or so, and the results are suggestive. Several ancient scribes mentioned that wild horses persisted in Iberia during the Classical period, but these claims disappear from the historical record around 2,000 years ago. By comparison, numerous writers attested to the presence of wild horses on the broad steppes of the Eurasian interior for the next 1,800 years or so. It is possible that these free-ranging equids were feral, but given that their locations correspond to the purported "refugia" where wild equids survived at the end of the Pleistocene, it seems more likely that references to "wild" horses really did document wild, never-domesticated equids, the close cousins from whom all domestic horses had recently diverged.[79]

Meanwhile, archaeological evidence suggests that selective regimes may have also intensified for dogs and pigs during the Classical period. For example, dogs had remained relatively uniform in size throughout most of European prehistory, but they began to manifest much greater morphological variation around 2,000 years ago. Lap dogs first appear in the archaeological record in Ancient Rome, and this diversity extended throughout the farthest reaches of the empire.[80] Furthermore, dogs remained essential weaponry, and special efforts were devoted toward cultivating their gene pool.[81] These genetically refined war dogs remained in the minority, however, and most dogs in Eurasia, Africa, and Australia remained free-ranging and genetically undifferentiated, and therefore akin to the modern semi-feral canids known as "pariah dogs" or "village dogs."[82]

Last but not least, pigs were also brought under more intense selective regimes during the Classical period. For example, historian Samuel White has shown that artificial selection was most intensive and most widespread during the Roman period, but that this "initial experiment in porcine development" ended when the Roman Empire fell to pieces.[83] The breakdown in intensive breeding programs meant that, for the next thousand years, swine farmers in Europe resumed the husbandry method known as pannage, herding their pigs into the woodlands where they subsisted on acorns and other mast. These pigs were feral, inasmuch as they relied on their wits for sustenance and procreated beyond humanity's watchful eye. Finally, conditions remained much different in China, where human population densities were much higher. Throughout the historical period, and in contrast to their European counterparts, Asian farmers kept their pigs confined in sties and fed them at troughs, which increased their propensity to fatten.[84] Despite their desirable qualities, however, these Asian pigs remained distinct from both European and African swine throughout the medieval period.

## Chapter Summary

Although scientists and scholars continue to debate the timing and location of the first domestication event, everyone can agree that dogs (*Canis lupus familiaris*) were the first animals to forge a domesticated relationship with humans, and that they first diverged from gray wolves somewhere in Paleolithic Eurasia. Scientists once assigned humans all of the credit for initiating the domestication of dogs, but many now favor the coevolutionary hypothesis, which acknowledges reciprocal influences between humans and canids. Meanwhile, recent genetic analyses confirm that the reproductive boundaries distinguishing wild, domestic, and feral canids were originally quite porous, and that the domestication of dogs was not achieved in a single generation.

Around 12,000 years ago, for reasons that remain unclear, humans living in the Near East (a region stretching from western Turkey to eastern Iran) gave up hunting and gathering and began settling into sedentary communities. During this Neolithic Revolution, humans entered into domesticated relationships with numerous plant and animal species. People and pigs (*Sus scrofa*) first linked up with one another in the Near East around 10,500 years ago, though similar unions were later forged elsewhere. Meanwhile, people and horses (*Equus caballus*) first joined forces on the steppes of northern Kazakhstan around 5,500 years ago, and soon diffused across the rest of Eurasia. Although dogs, pigs, and horses entered the domesticated relationship under vastly different pretenses, they all served vitally important roles during the first few millennia of European civilization.

By 3,500 years ago, dogs, pigs, and horses were distributed throughout most of Africa and Eurasia. Their movements were generally not restricted, and so they moved in and out of humanity's immediate sphere of selective dominion with surprising ease and frequency. Some dogs were subject to intense anthropogenic selection during the Roman Empire, but the vast majority remained genetically undifferentiated, and a great many continued to consort with wild, never-domesticated wolves. In most places, pigs retained the freedom to walk away from their relationship with humans at any time if they so desired, though there were exceptions to this pattern. In China, where human population densities were much higher, farmers penned their domestic pigs inside fenced enclosures from an early date. In Europe and Africa, the reproductive boundaries distinguishing wild, domestic, and feral pigs were actually quite porous, and there was significant admixture among all three populations. Finally, horses assumed subtly different morphologies as they spread into both southern and northern climes, though it is unclear if this diversification resulted from natural selection or anthropogenic selection. Domestic horses undoubtedly incorporated genes from wild stock, but, once again, it is not clear how often domestic horses left humanity's immediate selective dominion and contributed their feral genes to otherwise wild populations. In sum, the boundaries distinguishing wild, domestic, and feral animals remained rather ill-defined for thousands of years.

# 3

## WHEN FERALITY REIGNED

### ESTABLISHING AN OPEN RANGE IN THE COLONIAL SOUTH

*Here beyond men's judgments, all covenants were brittle.*
~ Cormac McCarthy

The vast majority of human history had already elapsed by the time humans discovered the Americas. Readers will recall that people had reached northeastern Siberia around 30,000 years ago, but that they were blocked from entering North America, or even knowing that it existed, by enormous walls of glacial ice. As a result, the first people to enter North America had to wait on the continental doorstep for thousands of years.[1] They braved harsh, tundra-like conditions, but at least they did not go it alone. Instead, both the genetic evidence and the archaeological record indicate that dogs lived among the people of Beringia during the late Pleistocene.[2] Both species remained isolated in this continental cul-de-sac for thousands of years, utterly unaware that more than 16-million square miles of *terra incognita* lay beyond the southeastern horizon.

People and dogs first entered North America, together, around 16,000 years ago, skirting the northwestern coast in small watercraft until they reached dry land.[3] They then diffused across the continent for the next several thousand years.[4] When they crossed east of the southern Appalachian Mountains, the region looked considerably different than it does now.[5] Ice sheets never extended into the South, but the region was still bitterly cold. The winters were longer and drier, and snow covered most of the landscape for most of the year. As a result of all these climatic changes, the actual shape of the South would have also looked much different. For example, the global ocean was about 400 feet lower than it is today, and so the Florida peninsula was nearly twice as wide. The region also contained no fewer than seventy-five different mammal species, a full third of which qualified as "megafauna" (heavier than 100 pounds). Among this extraordinary menagerie were bison, tapirs, peccaries, moose, caribou, ground sloths, mammoths, mastodons, grizzly bears, jaguars, dire wolves, and saber-toothed tigers. Many of these species disappeared from the zooarchaeological record soon after people and dogs arrived. As a result, some have suggested that the two species overhunted the region's faunal species and drove them extinct, while others insist that a warming climate was to blame.[6] Whether climate explains the Pleistocene extinctions or not, there is no doubt that global warming influenced the South in profound ways. Most significantly, the melting polar icecaps not only reshaped the southeastern coastline, but also re-established the distant Bering Strait between Siberia and Alaska. As a result, people and dogs in the Americas were isolated from their closest conspecifics on other continents.[7]

By 10,000 years ago, the biotic composition of the South assumed something resembling its present composition.[8] In relatively short order, mastodons, mammoths, and muskoxen gave way to squirrels, turkeys, and rabbits.[9] These replacement species may inspire yawns when compared to their Pleistocene predecessors, but remember that humans were infinitely more bizarre than saber-toothed cats or house-sized sloths. Numbering in the low thousands and clustering around water sources and stone quarries, the first humans in the South not only boasted advanced social coordination and deft use of tools, but they also came equipped with a companion species.[10] Acting as sentinels for one another and serving (presumably) as friends, people and dogs boasted something that no other mammals on the continent did: a tightly interwoven, highly advantageous, trans-specific partnership that helped ensure their mutual survival.

Because these people and their descendants (hereafter referred to as "Native Americans" or "Indians") did not write anything down, one has to rely on others kinds of evidence to learn anything about them or their dogs. For example, the oldest-known dog remains on the continent have been carbon-dated back to more than 11,500 years ago, and the zooarchaeological evidence suggests that dogs later diversified into countless regional varieties in the millennia thereafter.[11] In addition to scattered remains in midden heaps, zooarchaeologists have also discovered hundreds of dog burials scattered throughout the region.[12] Many of these dogs were buried with deliberate care after they died, which suggests a close bond in life. These burials date back to the Woodland Era (1000 BC to AD 1000), when people were still largely nomadic hunter-gatherers, as well as the Mississippi Era (AD 1000 to AD 1500), when they increasingly adopted maize-based agriculture and settled into communities.[13]

Remarkably, zooarchaeologists have uncovered more than a hundred dog burials from one Late Woodland site in southeastern Virginia, making it one of the largest pet cemeteries in the world. This assemblage also contains the skeleton of a woman interred with the skeleton of a dog curled at her feet, further evidence of the bond between two species.[14]

Scientists once believed that dogs might have also been independently domesticated from North American wolves, but recent genetic analyses have confirmed that dogs were originally domesticated in Eurasia, and only in Eurasia.[15] This means that ancient North American wolves did not contribute their genes to the dog's genome, but it does not preclude the possibility that dogs might have contributed *their* genes to ancient North American wolves.[16] Sure enough, geneticists have recently identified the melanistic mutation for black fur in canids. What is more, they have determined that the mutation first arose in dogs around 46,000 years ago, and that it later appeared in North American wolves. These scientists conclude that there must have been episodes of "introgression" after domestic dogs arrived on the continent. They might have just as easily called them episodes of "feralization." After all, their results prove that domestic dogs were not always subject to humanity's immediate selective dominion in the pre-contact era, and that they sometimes procreated beyond human control.[17]

Native American people and dogs remained genetically isolated from the rest of the world for thousands of years. That famously changed in 1492, when Christopher Columbus and a crew of more than eighty people plunged headlong across the potentially infinite ocean blue. There were no domestic animals on Columbus's first voyage, but there were plenty aboard his second voyage in 1493. These animals were turned loose throughout the Caribbean to live and reproduce on their own. In short order, the islands were overrun. Pigs proved especially fecund, but dogs and horses likewise established large feral populations. When Spanish conquistadors ventured to mainland Mexico, they continued to seed the landscape with feral animals, who once again reproduced in droves.[18]

One would think that the exact same pattern would have been repeated when Spanish explorers finally reached southeastern North America, and yet that was not the case.[19] They were surprisingly slow to even realize that a continent lurked nearby.[20] Several Spanish explorers had attempted to establish colonies in the South during the 1520s, and all of them failed. These attempts always included horses and dogs, and most of them included pigs, but there is no evidence that any of the animals established feral populations.[21] Hernando De Soto led the most extensive early foray into the region in 1539, when he landed an army containing more than 600 people, around 230 horses, 13 pigs, and an undetermined number of war dogs on the coast of southwest Florida. Once again, however, there is no definitive proof that the expedition resulted in any feral animals. Scholars once believed that the feral mustangs who roam the American West owed their origin to feral strays from the De Soto expedition, but that notion was disproved in 1938, when anthropologist Francis Haines used historical ecology to demonstrate that mustangs had first diffused northward from Mexico, beginning in the late seventeenth century.[22] By comparison, pigs seem more likely candidates for feralization, especially since the thirteen pigs whom De Soto originally brought had quickly swelled to several hundred.[23] It is entirely possible, if not probable, that some of pigs left the expedition and went feral, but the larger claim that they were the progenitors of today's extant population lacks proof.[24]

It was not until the English initiated colonization efforts in the South that the number of pigs, horses, and (non-native) dogs began to rise significantly. When English settlers first attempted to establish a permanent colony along the Outer Banks of North Carolina in the 1580s, they quite naturally took dogs, pigs, and horses with them.[25] Although these colonies failed to take root and the animals failed to establish feral populations, they are nevertheless instructive. Most significantly, these prospective colonists produced the first written description of native dogs in the South. To be fair, the De Soto expedition's chroniclers had also mentioned that Native Americans possessed "little dogs," but it was always in a culinary context.[26] English colonists likewise ate native dogs, but they also occasionally described them. For example, Thomas Harriot wrote that Native Americans in North Carolina possessed "wolfish dogs," while artist John White painted a watercolor showing a dog in an Indian village, the first graphic representation of a Native American dog by a European artist (Figure 2).[27]

The English renewed colonizing efforts in 1606, when three ships set sail for America under the direction of the Virginia Company. Like so many of their predecessors, these ships counted dogs among their number, but they did not contain any other domestic animals.[28] Nor did they pick up any animals when they passed through the Canary Islands or the West Indies, this despite the fact that one of the prospective colonists, George Percy, recalled seeing droves of "wild boars" on the islands.[29] They arrived in Tidewater Virginia during the spring of 1607, and

*When ferality reigned* 31

Figure 2 John White's watercolor of an Indian village, 1585 (detail)
An artist named John White was among the first English explorers to visit southeastern North America, and he later produced this watercolor depicting the village of Pomeiooc in eastern North Carolina. It is the earliest known visual representation of a Native American dog by a European.
**Source** – Paul Hutton and David Beers Quinn, *The American Drawings of John White, 1577–1590* (Chapel Hill: University of North Carolina Press, 1964).

quickly established a fort they named Jamestown on a river they named the James River. Their dogs wasted little time finding their way into the historical record. Soon after arrival, for example, John Smith gave the local paramount chief, a man known as Powhatan, a white dog as a gift and the chief purportedly cherished the animal.[30] Other records reveal that when Indians attacked an exploratory party led by Christopher Newport, the only casualty was an English dog.[31] Settlers also trained their canine companions to hunt and kill the large number of rats who had unexpectedly established a population in and around Jamestown.[32] Meanwhile, Smith provided a few comments on the Indians' dogs, whom he considered strikingly different from English dogs. "Their dogs of that country are like their wolves, and cannot bark but howl," he reported.[33] Smith's observation that native dogs resembled wolves echoed Harriot's description from a generation earlier, while his observation that they did not howl harkened Columbus's description of native dogs on Hispaniola more than a century earlier.[34]

Additional supply ships brought additional animals to the colony. The first pigs arrived in Jamestown in January 1608, and by the summer of 1609, there were more than sixty pigs in and around the fort.[35] Finding it increasingly difficult to manage so many pigs, the settlers transported the animals to a nearby island in the James River that they named "Hog Ile," or Hog Island.[36] This turned out to be a horrible blunder. Most importantly, the location left the pigs vulnerable to attacks when relations with the Indians turned sour. By the end of the year, Powhatan and his people had killed hundreds of pigs.[37] Meanwhile, the Virginia Company shipped the first eight horses from England in 1609.[38] While en route to Virginia, however, a terrifying hurricane scattered the fleet far and wide. The ship carrying the horses arrived in Jamestown relatively unscathed, but the rest of the supply contingent was still lost at sea. Almost all of their provisions had been aboard the flagship, the *Sea Venture*, the fate of which remained unknown.[39]

Dread befell the colony, and with good reason. By the end of 1609, food supplies had run out. The colony's few remaining dogs, horses, and pigs were ignominiously eaten during the "Starving Time" that winter.[40] When the *Sea Venture* finally arrived at Jamestown, its crewmembers discovered that 80 percent of the colony was dead, and that survivors were slowly starving. Surveying this horror, the crew resolved to go and retrieve food from Bermuda, where they had recently been shipwrecked, and where droves of feral pigs had roamed in abundance.[41] On their way down the James River, the settlers encountered Lord Delaware, who provided the woebegone colony with much needed provisions.

Conditions continued to improve throughout 1611. Thomas Dale arrived with additional pigs in May, while Thomas Gates brought still more pigs that August.[42] Meanwhile, horses continued to arrive on almost every ship thereafter. Recognizing that domestic creatures were critical to their success, colony leaders passed a series of laws that forbade anyone from killing any domestic animal upon penalty of death.[43] The prohibition had its desired effect. Although colonists had erected palisades across several peninsulas in an attempt to secure their livestock within enclosures, these fences proved ineffective.[44] By 1614, feral animals were seemingly everywhere. Ralph Hamor wrote that there were "infinite hogs in herds all over the woods," and that every person had "some mares, horses and colts."[45] In 1619, the Virginia Company confirmed that there were "some horses" and an "infinite number of swine broken out into the woods."[46] Two years later, another colonist confirmed that "horses and hogs are dispersed and grown wild in the woods."[47] Even though Virginia's feral animals had slipped beyond the colonists' control, the crown continued to claim ownership rights and legal dominion over royal "property." For example, when colonial authorities granted a colonist permission to take a hundred feral pigs out of the woods, it was with the understanding that he would repay the crown in due course.[48]

Not surprisingly, this rapid growth led to trouble with Native Americans. As evidence of these increased tensions, the English began instituting precautions. As early as 1618, Governor Argyll had issued an edict rendering it illegal to teach Indians how to hunt pigs, upon penalty of death.[49] When the first assembly of elected representatives in North America, known as the House of Burgesses, convened in Jamestown one year later, they reiterated Argyll's edict, and they further prohibited settlers from letting the Indians have any English dogs, like mastiffs, greyhounds, or bloodhounds.[50] The reasons behind this law were obvious enough. As tensions with the local Indians grew more strained, the settlers did not want to provide their enemies with their most effective weapons.

Given that an encroaching tide of feral animals preceded English colonists across the Virginia landscape, it is little surprise that Native Americans began targeting the four-legged transgressors. When they launched a large-scale attack that killed a quarter of all colonists in 1622, for example, Indians also targeted the colonists' animals.[51] William Rowlsley wrote to his brother that pigs had decreased precipitously during the conflict, and that feral dogs subsisted better than most colonists.[52] Not surprisingly, English settlers employed both dogs and horses in their attacks against the Indians. One colonist advised his compatriots that every house should keep a "good mastiff" to scare away Indians during the day, and to hunt them down during the night.[53] In similar fashion, another colonist suggested employing horses to chase the Indians, and bloodhounds or mastiffs to tear them apart.[54] These episodes are important, for they belie the "Black Legend" myth that ascribes the use of human-killing dogs to Spaniards exclusively.

The carnage took an enormous toll on the colony's livestock, and, as a result, authorities passed another resolution that prohibited people from killing domestic animals upon penalty of death.[55] Once more, the prohibition had its intended effect, and the number of animals soared throughout the 1620s. Pigs proved especially prolific, and they quickly wreaked havoc on the colonists' precious corn. The colonists originally responded by appointing keepers to herd the pigs, similar to the method of pannage used in medieval Europe, but they soon abandoned that practice in favor of fencing. In 1628, colonists constructed a palisade that stretched from the James River to the York River. They had hoped the palisade would turn the end of the Virginia Peninsula into an enclosure, but their marauding animals never obliged.[56] A year later, the House of Burgesses ordered every person in the colony to plant two acres of corn, instructing that the plots should be "sufficiently tended, weeded, and preserved from birds, hogs, cattle and other inconveniences."[57] When that dictate failed to prevent animals from trespassing, lawmakers directed all settlers to build fences that would enclose their crops.[58] This stipulation effectively gave livestock the run of the land, and thereby codified the "open range," a social, cultural, and ecological institution that would define life in the South for more than two hundred years.

Numerous sources attest to the ubiquity of feral pigs around mid-century. In 1649, William Bullock wrote that there were "an infinite number of hogs."[59] One year later, Edward Williams marveled that many Virginians profited from swine, "whose feeding costs them nothing."[60] In 1658, John Hammond repeated the claim that pigs were "innumerable" and that they were now being exported to New England and parts of the Caribbean.[61] In an attempt to keep track of who owned which pigs, colonists carved distinctive notches into the ears of their respective animals, and they registered these earmarks with the local courthouse.[62] This practice did little to dissuade poaching, though. Lawmakers passed a law to prevent rampant pig theft in 1647, and another in 1658. These measures evidently had little to no effect. In 1662, lawmakers acknowledged that "the stealing and killing of hogs is a crime usually committed and seldom or never detected or prosecuted."[63] Of course, most of the colony's pigs were born in the woods and bore no identifying insignia. The crown lay claim to these animals, and prohibited colonists from taking any from the forest. In light of the pig-saturated landscape of eastern Virginia, however, this prohibition smacked of absurdity.

By comparison, the number of horses had grown rather slowly during the first few decades of settlement. By mid-century, however, their numbers began to show a marked increase. In 1649, there were approximately 250 horses in Virginia, the majority of whom persisted in a feral condition.[64] As a result, the legislature passed several acts dealing with feral horses. In December 1656, complaints against horse stealing grew so loud that the Burgesses passed a law codifying the process whereby lost horses could be reclaimed. The law stated that anyone who took a horse from the woods was legally obligated to advertise their discovery with the county clerk. Conditions had evidently not improved five years later, when the colony passed a law stating that it was still a problem.[65] Making matters worse, feral horses often traipsed through the colonists' cornfields.

To try and mollify the situation, lawmakers employed a reliable method. In the past, colonists had successfully utilized domestic animals to help eradicate wild ones. They had long used dogs to track and kill wolves, but there were other examples. In 1632, lawmakers had stipulated that any colonist who killed a wolf would be rewarded with a pig.[66] In 1655, they had said that any Native American who killed a wolf would be rewarded with a cow.[67] When they tried a similar scheme with feral horses, however, it just did not take. In 1662, the colony passed a law stating that local authorities should tax horses, and use those funds to pay for a bounty on wolves.[68] The logic behind this decision was fuzzy, at best. After all, many of Virginia's horses lived in the woods, free from human stewardship, and colonists were unwilling to pay taxes on animals they rarely saw and whom they could just as easily pretend did not exist. The plan did not work, and lawmakers repealed the scheme three years later.[69]

By the late 1660s, the number of horses had grown dramatically. In 1668, lawmakers repealed a previous law that had forbidden colonists from exporting horses outside the colony.[70] One year later, they passed a law outlawing the further importation of horses, acknowledging that the number had grown burdensome of late.[71] One year after *that*, lawmakers updated their fence laws to account for all of the feral horses. To prevent the unruly animals from stomping through colonial cornfields, lawmakers mandated that all horses had to be enclosed from the middle of July until the end of October.[72] This statute was significant for a couple of different reasons. First, this was one of the earliest attempts to curb what had hitherto been an exclusively open range. Second, some believe that colonists first placed horses on Virginia's coastal islands at this time in an attempt to circumvent the new enclosure law, and that extant herds on Assateague Island trace their origin back to this original founding stock.[73] It may well be true that horses were introduced to Assateague at this time, but, as we shall see, the origin of the extant population is another matter entirely.

By 1670, there were about 8,000 horses in Virginia, and an "infinite" number of pigs.[74] This was a recipe for disaster. After all, the same conditions that had prompted Native Americans to attack colonists in 1622 never really abated. In 1631, Virginia had passed a law stating that "if any Indians do molest or offend hogs," the Indians would be considered "our irreconcilable enemies."[75] In 1646, a law had been passed granting colonists permission to pursue pigs on the north side of the York River, which was still technically Native American land.[76] In the decades thereafter, the colony tried to mollify the situation with legislation. In 1662, lawmakers acknowledged that feral animals were often to blame for antagonistic encounters between Indians and colonists. Accordingly, they passed a law requiring "Indians to fence in a cornfield proportional to the number of persons the said Indian town doth consist of."[77] As Virginia's lawmakers saw it, the problem was not the plague of feral animals, but rather the fact that Indians did not erect fences to keep said animals out. Not surprisingly, the colony's solution solved nothing. In 1663, Edmund Scarburgh reported that Indians continued to steal pigs from the forest on a daily basis.[78] In 1674, lawmakers reiterated that Indians should fence their corn, and they dictated that Indian pigs must bear some kind of mark denoting legal ownership.[79]

All too frequently, these conflicts over feral animals erupted in violence. In the summer of 1675, several Doeg Indians ventured south of the Potomac River to take pigs from colonists whom they felt had not bartered in good faith.[80] Colonists pursued the Indians, found them, and reclaimed the pigs. The Indians, in turn, killed several colonists.[81] Passions were inflamed on both sides. Nathaniel Bacon, who lived on the frontier, riled up support along the margins of the colony. One of the reasons he was able to do so was that now there were enough horses to mount the previously marginalized, and pedestrian, frontiersmen (by this point, there was one horse for every five colonists). This favorable ratio allowed Bacon to conjure an army that consisted of more than 700 horses, which allowed his forces to cover more territory.[82] He had 120 horses with him when he forced Governor Berkeley (at gunpoint) to grant him permission to wage war against Indians.[83] Not all of Bacon's horses were obtained by honest means. One petition suggests that Bacon forcibly impressed many of these animals from his neighbors, though the colony's commissioners believed that he conscripted most of his horses from the woods.[84] Under threats of physical harm, the Burgesses enacted several laws during the 1676 session meant to address the frontier conditions, especially as they pertained to feral animals.[85]

By the late seventeenth century, it was clear that Virginia's pigs and horses faced subtly different selective pressures. The vast majority of pigs lived in a feral condition, and, as such, visitors were "astonished" by their apparent ubiquity.[86] Even people who were born and raised in Virginia offered similar testimony. "Hogs swarm like vermin upon the earth," Robert Beverley acknowledged, adding that the animals typically run in "a gang" and that they "find their own support in the woods, without any care of the owner."[87] Meanwhile, the number of horses also continued to grow. Most of these animals were born in the woods, had never known human companionship, and were thus "as shy as any savage creature."[88] Every spring, some of the colonists hunted feral horses for sport. They constructed pitfalls near the animals' watering holes, and one Swiss visitor confirmed that "their meat is good to eat."[89] Others had no interest in killing the colony's feral horses, and instead drafted the animals into service on an as-needed basis. One visiting Frenchman noted that colonists employed horses for every sort of travel, even if it was just a few hundred yards down the road. Horses were so ubiquitous, he wrote, colonists typically just surrendered the animals back to the woods after they reached their destination.[90]

Virginia's colonists continued pushing deeper into Native American territory throughout the eighteenth century, but they were no longer the only ones doing so.[91] Several hundred miles down the coast, another English colony had also successfully attached itself to the southeastern substrate. In 1670, 200 colonists, mostly Barbadians of English descent, established Charles Town (later renamed Charleston) near the mouth of what is now known as the Ashley River.[92] Like their Virginia brethren, Carolina's colonists found it easier to fence their crops and to let their animals roam free. As a result, feral animals were everywhere in relatively short order. By the 1680s, pigs were so abundant in Carolina that even the poorest colonists could afford to own a couple hundred. These pigs provided protein, but they also provided a commodity that could be exported to England and the West Indies.[93] Carolina's colonists tried to mark their pigs' ears with distinguishing notches, just as they had in Virginia, but most of Carolina's pigs nevertheless went unmarked and unclaimed, just as they did in Virginia.[94] Meanwhile, the horses in Carolina increased at a slower rate than pigs, but they increased nonetheless. By 1682, there were about 150 breeding mares in Carolina, many of them imported from New York and Rhode Island.[95] Carolina lawmakers passed several laws addressing feral animals throughout the 1690s, including no fewer than three laws encouraging the destruction of "wild and unmarked" livestock.[96] In 1694, they passed a law mandating that fences around crops should be 6 feet tall.[97]

It was not just Europeans who moved into Carolina, however. Owing to the horrifying transatlantic slave trade, a growing number of people who had been violently uprooted from their homes in western Africa also came to populate the colony. A small number of Africans had accompanied Spanish explorers to the South during the sixteenth century, but there were still relatively few displaced Africans in Virginia as late as 1681 (just 3,000 out of an estimated 80,000 people in the colony).[98] The settlement process looked considerably different in Carolina, where enslaved Africans had accompanied the very first settlers.[99] By 1708, a majority of non-Native American people living in the colony (5,500 out of 9,500) were enslaved Africans.[100] Most of these people had lived in the Bight of Benin, the Gold Coast, or the Bight of Biafra prior to their abduction, though nearly all of them passed through slave markets in other English colonies, including Barbados and Jamaica, before their arrival in Carolina.[101] Many of them had worked with livestock in their homeland, and were compulsorily tasked with doing the same in Carolina.[102] This remained true even after rice became the leading agricultural pursuit in the 1710s, and livestock roamed deeper into the hinterlands. Since the enslaved were charged with rounding up wayward cattle, it is no exaggeration to say that many of America's earliest cowboys were recently displaced Africans.[103]

According to some historians, Carolina's earliest enslaved Africans experienced a comparatively large measure of autonomy during the early colonial period. During so-called "free time" at the end of a hard day's work, slaves were "permitted" to tend their own crops and range their own animals on the vast commons. To be sure, many colonists objected to this practice, insisting that enslaved people should not be allowed to own animals. In 1714, the legislature passed a law that forbade slaves from owning pigs or horses, though it should be noted that the prohibition proved nearly impossible to enforce.[104] After all, the enslaved often knew the Carolina landscape more intimately than did their owners.[105]

It was not long before the colony's proprietors officially recognized that the northern portion of Carolina was largely distinct from the southern portion. Its settlement history was certainly different. Colonists from Tidewater Virginia had first settled in the Albemarle Sound region as early as the 1660s, while others trickled northward from Charleston throughout the 1680s.[106] The Lords Proprietor first established North Carolina as a distinct government in 1691, and provided the territory with its own governor in 1712.[107] Following the lead of lawmakers in Virginia and South Carolina, the North Carolina legislature passed a law articulating "what fences are sufficient" in 1715. Once again, the law privileged the rights of stockowners over the rights of planters, who were responsible for erecting a fence at least 5 feet tall around their fields to protect them against the scourge of "horses, hogs or cattle" that plagued the young colony.[108]

This radiative method of free-range husbandry led to significant conflict between Europeans and Native Americans in North Carolina and South Carolina. The Nottoway Indians of Albemarle Sound had clashed with English settlers regarding the rightful ownership of local pigs as early as the 1660s, and they were still struggling to resolve the issue thirty years later. In 1699, colonial authorities mandated that if the Nottoway wanted to claim pigs as their own, then they must physically mark the creatures in some way.[109] In 1705, colonists accused Indians of stealing pigs.[110] In 1707, the North Carolina Governor's Council complained that Indians planted corn without fences, "so that no English can seat near them without danger of trespassing by their cattle and horses."[111] Four years after that, the governor of North Carolina, Edward Hyde, wrote a letter to his counterpart in Virginia, Alexander Spotswood, complaining that Indians in Virginia frequently targeted free-ranging animals who technically belonged to North Carolina colonists.[112] Finally, but perhaps most significantly, feral animals also served as vectors for a variety of zoonotic diseases, including smallpox, that devastated Native American populations in both North Carolina and South Carolina.[113]

Given that the Indians of North Carolina felt their very existence threatened, it is little surprise that they eventually stopped killing European animals and started killing European people.[114] During a single attack in 1711, Tuscarora Indians killed more than a hundred colonists and an untold number of animals, both domestic and feral, scattered along the Neuse River.[115] Those who survived the attacks understood that the loss of their animals left them vulnerable, and they filed a petition complaining that Indians continued to kill their horses on a daily basis.[116] One year later, Thomas Pollock wrote a letter to the Lords Proprietors complaining that Indians had "carried away" their animals, and that there was "little or no pork" on hand.[117] It was not just the colonists' animals who suffered, though. By this point, many (but not all) groups of Indians in the Carolinas had adopted European dogs, pigs, and horses, and some assisted the English in the eradication of their Native American cousins. What is more, their collaboration with the colonists meant that their own animals were sometimes the target of violence. In 1714, the Chowan Indians protested that they had joined the English colonists on eight expeditions "against the Indian enemy of this province," and that they had lost seventy-five pigs and two horses as a result.[118]

Meanwhile, just as colonists in North Carolina followed their feral animals into the Tuscarora War, colonists in South Carolina followed theirs into the Yamassee War. Sounding familiar protests, the Yamassee Indians complained time and time again throughout the first decades of the eighteenth century that the colonists' free-roaming horses and pigs ransacked their cornfields. Furthermore, those same animals outcompeted and drove away the native deer, whose skins the Yamassee collected for their trading value.[119] Finding their lives threatened by this ecological crunch, to say nothing of the sprawling, intraregional Indian slave trade that devastated their numbers, the Yamassee launched a series of bloody attacks that claimed the lives of hundreds of people and innumerable animals.[120] In response, colonists once again directed their dogs and horses to attack Native Americans.[121] Thousands of Yamassee Indians died in the violence that followed. Survivors fled inland toward the mountains or southward toward Florida, while others settled in tributary communities near European settlements.[122]

As colonists in the Carolinas killed or uprooted local Native Americans, feral animals moved in to replace them. As might be expected, pigs were particularly successful. Soon after John Lawson arrived in Carolina in 1700, he remarked that huge droves of pigs roamed the colony, and that the suids' steady diet of acorns, nuts, and fallen fruit had produced "some of the sweetest meat that the world affords." Fans of Carolina barbecue will no doubt second his opinion that "the pork exceeds any in Europe."[123] English naturalist Mark Catesby, who was one of the first scientifically inclined people to describe the flora and fauna of southeastern North America, agreed with Lawson that the mast-fed pigs of North and South Carolina provided the best-tasting pork on Earth. "Their flesh excels any of the kind in Europe," he confirmed.[124] Not everyone saw this as a virtue, though. William Byrd of Virginia (who never missed an opportunity to disparage his neighbors to the south) wrote that "the inhabitants of North Carolina devour so much swine's flesh that it fills them full of gross humors."[125] Byrd's insult was not entirely fair, though. After all, many of the pigs who were born in North Carolina ended their lives elsewhere. In 1727, for example, an estimated 30,000 pigs were driven out of North Carolina and into Virginia.[126] In 1733, members of the North Carolina General Assembly estimated that 50,000 pigs were driven into Virginia, a 66 percent increase in just six years.[127]

Meanwhile, a majority of horses in the Carolinas remained feral throughout the colonial period. That does not mean that people went without horses, though. On the contrary, just about every person in the colony could lay claim to a horse. When Jean Pierre Furry visited in 1731, he noticed that hardly anyone traveled by foot, and that even enslaved Africans traveled by horseback.[128] Meanwhile, the emerging Creek and Cherokee confederacies had both long since obtained horses by the 1730s.[129] Despite these facts, feral horses continued to outnumber domestic ones. John Brickell reported great droves of feral horses roaming the savannas in the 1740s, noting that the horses often roamed more than fifty miles from the closest human.[130] Perhaps not surprisingly given this general lack of interest in husbandry, thievery plagued the Carolinas throughout the eighteenth century. Lawmakers in South Carolina passed several laws to discourage thievery, but the practice remained widespread throughout the colonial period.[131] In 1752, visitors to North Carolina confirmed that "nefarious" horse thieves abounded in certain parts of the state.[132]

Many of the colonies' horses could live their entire lives without ever seeing a person, but even so they were still not *entirely* free from anthropogenic influence. Ever since the late seventeenth century, lawmakers had encouraged colonists to let bigger horses go at large, and to eliminate smaller ones. Virginia first codified this practice in 1686, when legislators determined that quantity was not the same as quality. That year, the colony passed an act for "better improving" the breed of horses, which prohibited horses beneath a height of 13½ hands (approximately 52 inches) from roaming the commons.[133] This law was significant, for it marked the first time that any government in the South had attempted to direct the evolution of the feral horses under its charge. The measure evidently had little effect, however, because the House of Burgesses passed identical laws in 1713 and 1748, both of which prohibited horses under 13½ hands from roaming at large.[134] Yet evidence from the Carolinas suggests that the prescriptive measures sometimes worked as they were intended. For example, when North Carolina's lawmakers first attempted to shape *their* colony's feral horses in 1723, they stipulated that horses less than 13 hands tall would not be allowed on the range. By 1768, they had amended the law to prohibit horses under 14 hands (56 inches) from roaming at large.[135]

Nor was that the only indication that colonists were interested in shaping their horses' collective gene pool. When he passed through Virginia in the late 1750s, English clergyman Andrew Burnaby observed that the "gentlemen of Virginia, who are exceedingly fond of horse-racing, have spared no expense or trouble to improve the breed of them by importing great numbers from England." British traveler J. F. D. Smyth offered similar testimony when he visited Virginia in the early 1770s. "The Virginians, of all ranks and denominations, are excessively fond of horses, and especially those of the race breed."[136] Charleston's earliest recorded horse race took place in 1734, the same year that wealthy South Carolina planters founded the Carolina Jockey Club. In the 1740s, planters began importing Arabian horses for racing and breeding.[137] Between 1764 and 1766, racers in the Chesapeake imported twenty-nine "thoroughbred" horses from Europe.[138] When Brickell visited North Carolina, he noted that there was a racetrack near every town.[139] Colonists were especially fond of "quarter-racing," in which two horses raced for a quarter-mile. Smyth dared suggest that no horse on Earth could out-run the "quarter-horses" of the American South.[140] Whether true or not, these comments are evolutionarily significant, for they signal the earliest evidence of intensifying anthropogenic selection in the region.[141]

Though we have as yet paid relatively scant attention to the colonists' dogs, that lack of attention does not mean that they were not present. On the contrary, their very ubiquity rendered them inconspicuous. As early as 1650, a Virginian named Edward Williams had advised his fellow colonists to keep "half a dozen good dogs" to scare away Native Americans.[142] In 1688, a parson in Jamestown had noted that "every house keeps three or four mongrel dogs."[143] Other sources also hint at the dog's ubiquity. In 1754, the Catawba Indians complained that European dogs acted as periphery sentinels, alerting their owners whenever Indians drew near.[144] Meanwhile, during the 1760s, an Anglican minister in South Carolina complained that a gang of "rude fellows" brought more than fifty dogs to town and set them to fight during church services.[145] Despite the dogs' widespread diffusion, colonial governments nevertheless sought to limit who could own the animals. In 1729, lawmakers in North Carolina passed a law that prohibited enslaved people from hunting with dogs, "unless there be a white man in his company."[146] In 1752, Virginia passed a similar law that likewise prohibited enslaved people from walking with dogs when traveling between plantations.[147] These rules were virtually impossible to enforce, however, and many slaves continued to own dogs throughout the colonial period.

European colonists and African slaves were not the only ones who kept dogs, though. In fact, numerous observers described the Native Americans' dogs. Not unlike Thomas Harriot and John Smith more than a century earlier, these colonists almost always compared the native dogs to wolves. During his visit in 1714, Lawson remarked that native dogs "are seemingly wolves made tame with starving and beating," and he lamented that Indians were "the worst dog masters in the world." He claimed (incorrectly) that Indians had no dogs before Europeans arrived in the South, and he was not alone.[148] Catesby made the same incorrect claim when he visited in the 1720s. What is more, he drew no distinction between wolves and native dogs, insisting that wolves were "domestic with the Indians."[149] Brickell offered similar observations during his visit to the Carolinas, writing that native dogs "are like the wolves in these parts." Echoing Lawson's sentiment (suspiciously verbatim), Brickell reported that Native Americans were "the worst dog masters in the world." Unlike Beverley, who had described at great length how Virginia Indians would use dogs when hunting, Brickell insisted that Carolina's Indians never used dogs when hunting.[150]

Just as a new colony had emerged to the south of Virginia sixty-three years after the founding of Jamestown, a new colony emerged to the south of Carolina sixty-two years after the founding of Charleston. In 1732, more than a hundred prospective colonists left England en route for the disputed expanse between British South Carolina and Spanish Florida. After they had arrived on the southeastern seaboard, but before they had even picked a spot to settle, they sailed to Charleston harbor so they could procure certain necessities, including pigs.[151] Only then did they head south, eventually establishing the town of Savannah on some bluffs near the mouth of the Savannah River. There were few locals with whom to contend, because most had fled south or been slaughtered following the Yamassee War.[152] In March, 1734, these settlers welcomed a group of Lutherans who were fleeing persecution in their native Salzburg, Austria.[153] These settlers, known to Georgia historians as the Salzburgers, disembarked twenty-five miles upriver from Savannah, where they set about building the town of Ebenezer. Though the animals within these two settlements were practically within shouting distance of one another, they had quite different experiences.

The colonists at Savannah built no enclosures for their animals, and their pigs therefore escaped to the woods beyond town almost immediately.[154] Even though the fecund pigs multiplied in the years immediately thereafter, colonists could not count on them for reliable sustenance.

"The killing and stealing of hogs has been so frequent at Savannah that there is hardly one person in that town that has one," the colony's founder, James Oglethorpe, complained in 1739.[155] Other colonists voiced similar grievances. "I have had very bad luck," William Ewen lamented in 1741. "Many of my hogs ran away, which kept me poor, for I could get none to sell."[156] That same year, a colonist named Samuel Perkins decided to pull up stakes and head back to Europe, explaining that he had once boasted "a very good stock of hogs," but that they had all been killed, stolen, or lost while roaming the woods.[157] In 1743, yet another colonist, Thomas Upton, complained that Indians had stolen pigs from his fledgling plantation.[158] He insisted that plenty of pigs had once roamed the range near Savannah, but that they had all been gathered up and shipped to other colonies.[159] Yet when Edward Kimmer visited coastal Georgia that same year, he observed numerous feral pigs.[160]

Meanwhile, colonists devised different methods for controlling their horses. Many of the animals were placed on Sea Islands off the coast of Georgia, where they were allowed to roam and reproduce without human supervision. Oglethorpe had personally placed horses on both Cumberland Island and Amelia Island.[161] Being marooned on islands may have allowed Georgia's population of feral horses to grow, but it also made them easy targets for aggrieved enemies, and many of them were killed when Spain attempted to invade the colony in 1742.[162] Not all of the island horses were killed, though, and there were still at least eight horses on Jekyll Island in 1746.[163] Those who did not own or have access to islands were forced to range their horses in the woods, where the animals subsisted on almost nothing.[164] Because these horses were not fenced and rarely seen, however, they were often subject to theft.[165]

Not all of the animals in colonial Georgia were encouraged to go feral, though. Things were different in the Salzburger colony of Ebenezer. Whereas most of the earliest British settlers had been urban-dwelling tradesmen and artisans before moving to Georgia, most of the Salzburgers had been farmers.[166] Thus, while Savannah's colonists turned their livestock loose in the woods, the Salzburgers provided their animals with food and shelter every night.[167] In 1745, the Salzburgers' spiritual leader, John Martin Bolzius, recorded with great excitement that he had received a copy of Jethro Tull's *Horse Hoeing Husbandry*.[168] Two years later, most of Ebenezer's horses were under yoke.[169] Despite the Salzburgers' best intentions, however, practicing good husbandry was not easy when so many of Savannah's feral creatures lurked about, and Bolzius sometimes complained that the British settlers' feral animals lured away the Salzburgers' domestic ones.[170] In fact, the Salzburgers were sometimes aroused to help their neighbors from Savannah track down their feral creatures.[171]

Originally founded by a group of trustees, Georgia became a crown colony in 1752.[172] Almost immediately, lawmakers passed legislation granting domestic animals free rein over the commons. In 1755, the new government passed a fence law that required colonists to fence in their crops and fence out their animals.[173] If any accidents befell said animals, planters would be charged with failure to erect a proper fence. Lawmakers passed another fence law in 1759, noting that the act from four years before had proved "very ineffectual" against feral animals.[174] Indeed, though Georgia colonists marked their livestock with ear notches, most of the colony's animals bore no marks. What is more, life in the woods had rendered them incredibly tenacious. One visitor from Germany noted that Georgia's pigs were behaviorally quite wild and that they wreaked constant havoc. "They break through the fences, swim through the water, and destroy everything," he wrote.[175]

Much as it had in Virginia and both Carolinas, this method of feral husbandry led to increased conflict with the Native Americans in Georgia. Significantly, however, the experience was slightly different than it had been farther north. After all, the Powhatan Indians of early-seventeenth-century Virginia had had no horses of their own, but the Creek Indians of mid-eighteenth-century Georgia had long since obtained horses for themselves. What is more, they obtained them from different places. They collected horses from the droves that surrounded European settlements along the southern Atlantic coast, they raided Apalachee settlements along the Gulf coast, and they traded with their western neighbors, the Choctaw, who first obtained horses from the West by 1690.[176] Perhaps not surprisingly, colonists and Indians routinely accused one another of stealing horses.[177] Meanwhile, despite their intense interest in horses, the Creeks had no interest in pigs or other livestock. They were still firmly entrenched in the deerskin trade during the colonial period, and while horses facilitated the transport of deerskin, pigs competed with deer for mast. As a result, Creek Indians were far more likely to kill pigs than to steal them.[178]

As was the case everywhere, the rising number of humans in Georgia also meant more dogs. Significantly, this was true of both Creeks and colonists. When Dutch cartographer Bernard Romans surveyed Georgia and Florida in the 1760s, he noted that Native Americans in Florida, near present-day Pensacola, kept entirely too many dogs. "They, like all other savages, are very fond of dogs, insomuch as never to kill one out of a litter," he lamented. He also reported, with some measure of exaggeration one is inclined to believe, that Indian dogs were so scrawny and pathetic that they had to lean up against a wall or a post before venturing to bark.[179] Meanwhile, colonists complained that there were far too many dogs clogging the streets of Savannah and other public roadways, and that the free-roaming canids had "become a very great nuisance." Lawmakers encouraged colonists to destroy any dog who was found "loose and at large." Even pet dogs were not safe. In 1766, they explicitly stipulated that if any dog bit, harassed, or even barked at a human, that animal could be hastily confiscated and the animal's human owner would be charged ten shillings sterling.[180]

Finally, even though Florida was the first part of the South that Europeans visited, the peninsula's settlement history was obviously much different from the rest of the region. Despite more than two centuries of residency, there were still relatively few Spanish living in Florida by the mid-eighteenth century.[181] In the early 1760s, Florida's largest city, St. Augustine, contained scarcely 3,000 souls, and the zooarchaeological record shows that pigs played a surprisingly small part of their diet.[182] Following the end of the French and Indian War in 1763, Spain surrendered La Florida to Britain, and most of the Spanish left. British troops arrived in July, and it was not long before livestock herders from South Carolina and Georgia came trickling south. They were not the only ones migrating into Florida at the time. Many of the state's first proto-Seminoles began moving into north-central Florida around 1763, and almost all of them migrated alongside dogs, pigs, and horses.[183] This fluid category included no small number of Africans who had escaped more tyrannical bondage to the north.[184]

William Bartram provided the best descriptions of the Seminoles' livestock management practices when he visited Florida just prior to the American Revolution.[185] He first arrived in northeast Florida in the spring of 1774, when he joined up with several British traders at a prearranged rendezvous on the St. Johns River. The party then traveled on horseback toward the Alachua Savannah to trade with Seminoles. Entering the savannah, Bartram observed large herds of "Seminole horses" coursing over the plains. He felt that they were "the most beautiful and sprightly species" of horse that he had ever seen (Figure 3). Like Catesby, he claimed that these animals were descended from Andalusian horses whom the Spaniards had turned loose centuries earlier. And, like Catesby, he viewed the population with a naturalist's eye. He traveled

Figure 3 William Bartram's sketch of the Great Alachua Savanna, 1765 (detail)
William Bartram spent considerable time in the Alachua Savanna (also known as Paynes Prairie) when he accompanied horse traders across sparsely settled parts of northern Florida in the late eighteenth century. He reported large droves of feral horses on the savanna, and he even sketched some of them. In this image, a feral horse can be seen just right of center.
**Source** – William Bartram, *Travels through North and South Carolina, Georgia, East and West Florida*, Philadelphia: James & Johnson, 1791). American Philosophical Society, Philadelphia, Pennsylvania.

extensively throughout the South, and he noticed that the animals looked different in different places. "I have observed the horses and other animals in the high hilly country of Carolina, Georgia, Virginia, and all along our shores (and they) are of a much larger and stronger make than those which are bred in the flat country next the sea coast," Bartram asserted. He asked as "a matter of conjecture" whether the environment somehow influenced the animals' morphology. He cited similar examples among other species, including the fact that a deerskin obtained from the Cherokees (who lived in mountains) weighed twice as much as one obtained from Seminoles (who lived in the lowlands).[186]

Throughout the summer, Bartram accompanied the traders as they scoured the Florida wilds in search of feral horses. European and Native American traders interacted and cooperated with one another extensively. Bartram moved easily in this dynamic world, traveling all over north Florida rounding up free-ranging horses. As he quickly found out, however, herding feral horses was no easy task. Many of the younger horses insisted on running away, which sometimes made the going slow. After the traders made camp, they invariably set about trying to subdue their feral bounty, "breaking, and tutoring the young steeds to their duty." On one occasion, Bartram happened across a party of more than forty Seminoles who had just sold a large herd of horses in St. Augustine, and had "furnished themselves with a very liberal supply of spirituous liquors" with the proceeds.[187] On another occasion, he described a frontier cow-pen where four or five "negroes," the aforementioned black cowboys of southern lore, retrieved, traded, and otherwise controlled large numbers of semi-feral horses.[188]

No less significant, Bartram also described Florida's canids. He recounted seeing a Seminole Indian in northern Florida who had trained his dog to help him keep track of a band of horses. This particular dog stayed out on the range with the horses and made sure none of them strayed away. Ever the naturalist, Bartram noted the animal's habitat, morphology, and ecology. He reported that the dog trekked home for meals, but never slept in or around the house at night. He further observed that the Seminoles' dogs were almost always black, and that they looked identical to the local wolves. Only their behavior and the fact that they could bark set them apart. He also repeated claims that some of the local wolves were piebald, which could suggest considerable admixture with feral strains.[189] Indeed, Bartram had plenty of opportunity to observe wolves, who were, after all, the dogs' closest wild, never-domesticated cousins. When he and his father first toured northern Florida in 1766, they heard numerous wolves along the St. Johns River. He elsewhere reported that wolves were still "numerous enough" in Georgia.[190] When he returned to Florida in the 1770s, he described the local wolves in detail. "The wolves of Florida are larger than a dog, and are perfectly black, except the females, which have a white spot on the breast," he wrote in his journal, adding that that the wolves in the

South were not as large as the wolves in Pennsylvania or Canada. Finally, he noted that on colder winter nights, the wolves of Florida would assemble in companies and howl throughout the night, "which is terrifying to the wandering bewildered traveler."[191]

It should be noted that the wolves whom Bartram observed were part of a "retreating edge."[192] Wolves were once abundant throughout the South, but colonial bounties hastened their demise. In the early seventeenth century, the House of Burgesses in Virginia passed numerous laws that were designed to eradicate wolves, and, by the late seventeenth century, wolves no longer troubled the most populated parts of the colony. When Durand of Dauphine visited eastern Virginia in 1686, he reported that wolves had grown "quite scarce."[193] In 1724, clergyman Hugh Jones of Williamsburg confirmed that "the wolves of late are much destroyed."[194] The other colonies also passed legislation to destroy the wolves in their vicinity, and, as a result, the region's wolf population declined precipitously during the colonial era. According to historian Jack Temple Kirby, wolves had all but disappeared from the South by the early eighteenth century.[195]

Indeed, by the end of the colonial era, the region's human population also looked considerably different than it had just one century earlier. In 1685, approximately 100,000 people lived south of the Potomac River. By 1775, there were more than 900,000. A majority of these people (around 527,700) were descended from Europeans. Many were descended from English settlers, but many were not. In particular, huge numbers of people from Scotland and Ireland began moving into the region around 1715. Many filtered into the Shenandoah Valley from more densely settled Pennsylvania, while others arrived in port cities like Charleston and Savannah.[196] Meanwhile, the region's Native American population had been reduced to a fraction of its former size.[197] Although hundreds of thousands of Native Americans had lived in the region south of the Potomac River prior to contact, fewer than 53,000 of them remained in 1685.[198] There were many reasons for this devastating collapse, but the three chief culprits were European disease, European technology, and European behavior.[199] Entire nations were wiped out, and though survivors from across the South and beyond converged to form new confederations, there were fewer than 17,000 Native Americans left by 1775.[200] The vast majority, around 14,000, lived in the Georgia backcountry, while scarcely 200 lived in Virginia.[201]

The shrinking number of Native Americans meant that southern colonists were forced to look elsewhere for slaves. After the Yamassee

War, for example, they imported and enslaved an enormous number of African men and women. By the end of the colonial period, Virginia had imported more than 77,000 Africans, while South Carolina had imported more than 110,000.[202] The vast majority of slaves who arrived in South Carolina prior to the American Revolution were imported directly from Africa, while smaller markets generally obtained slaves from Caribbean traders.[203] By 1775, there were approximately 364,000 people of African descent living in the South. Half of them lived in Virginia, though their presence was perhaps most conspicuous in South Carolina, where they drastically outnumbered people of European descent.[204] Despite colonial efforts to the contrary, many enslaved people laid claim to their own dogs, pigs, and horses throughout the colonial period.[205]

*Chapter Summary*

Migrating into the South necessarily influenced the geographic range of dogs, pigs, and horses. Dogs first migrated into the region alongside people thousands of years ago, and they were joined by pigs and horses (and more dogs) following European contact. Meanwhile, after several failed attempts in the late sixteenth century, English colonists successfully established their first toehold in the South when they founded Jamestown in 1607. Rather than waste precious time, labor, and materials building fences to enclose their livestock, these colonists allowed their animals to roam the landscape as they pleased. Free from direct anthropogenic oversight, pigs and horses began to manifest significantly different morphological and behavioral traits. They grew tougher and more tenacious as natural selection winnowed the weakest among them away. When additional settlers established additional colonies in Carolina, Georgia, and Florida, they too almost always provided their pigs and horses with unencumbered access to the land. This was the origin of the southern range, a social, cultural, and ecological institution that would define life in the region for hundreds of years.

During the seventeenth and eighteenth centuries, colonial husbandry was radiative in nature, and droves of pigs and horses spread across the landscape. In doing so, they invariably traipsed across Native American territory. Throughout the colonial era, some of these animals remained close to colonial habitats, where they could be called back into service on an as-needed basis, but a much larger number escaped to the woods. Pigs braved natural selection with a vengeance, and their legendary fecundity helped them establish thriving populations. Horses were likewise encouraged to live in a feral condition, but they were also more likely to be employed in domestic service by colonists, slaves, or Native Americans. Without question, the most intensive anthropogenic selection during the colonial period was directed not toward agriculture but rather toward developing faster horses for the purpose of racing. It should be noted that not all animals were allowed free rein over the landscape, and that ferality was not inevitable. A few populations were even enclosed from the very beginning.

Finally, the colonial experience also influenced the way each species engaged with humans. Living in an almost exclusively feral condition, pigs and horses forgot all their social graces. They developed "unruly" behavior to accompany their unkempt appearance. Meanwhile, there were two distinct populations of dogs in the colonial South, those of Native American descent and those of European descent. Both populations were genetically undifferentiated and their movements were not restricted, and yet, despite this relaxed anthropogenic selection, dogs generally retained close behavioral ties with their human commensals. In sum, the selective parameters distinguishing domestic and feral animals were not well defined during the colonial period.

# 4

## NASCENT DOMESTICATION INITIATIVES AND THEIR EFFECTS ON FERALITY

### CLAIMING DOMINION IN THE ANTEBELLUM SOUTH

*Control can never be a means to any practical end ...*
*Control can never be a means to anything but more control.*
~ William S. Burroughs

When the fledgling United States conducted its first census in 1790, officials counted more than 1,400,000 people of either European or African descent living south of the Potomac River. They did not bother to count the region's dwindling number of Native Americans, though historians now estimate that there were still around 25,000 of them living between northern Virginia and southern Florida in 1790.[1] Conditions had changed but little in 1803, when Benjamin Smith Barton provided the first scientific analysis of Native American dogs.[2] He reported that people in North America and South America had developed distinct breeds prior to contact, citing the ill-fated mute dogs whom Columbus first encountered on Hispaniola as evidence of anthropogenic selection. He wrote that mute dogs were not kept as pets in southeastern North America, apparently unaware that two centuries earlier, in Virginia, John Smith had reported dogs who could not bark.[3] Barton wrote that the dogs who lived among early-nineteenth-century Cherokees were "of a more mixed breed." As he explained it, this resulted from high levels of historical interaction between colonists and Cherokees. "The Cherokees themselves are so much mixed with the Europeans that they are often named by the traders, the 'Breeds,' " he wrote. Despite this widespread genetic admixture, however, Barton felt confident that European dogs had not completely subsumed Native American dogs, and that unadulterated pockets probably persisted somewhere on the continent. "It is likely that when seen he has been sometimes mistaken for the wolf," he reasoned.[4]

Meanwhile, the region's Native Americans accrued more and more pigs. One census from 1809 counted approximately 13,000 Cherokee people, but more than 19,000 Cherokee pigs. When officials conducted another census in 1826, they counted 18,000 Cherokee people, and more than 46,000 Cherokee pigs.[5] Farther south, Native Americans had slightly different experiences. The Creeks in Georgia had always been loath to adopt pigs, who competed with deer for mast. In 1796, Creek representatives complained to the young US government that their hunting grounds were "full of cattle, hogs, and horses."[6] As the federal government laid claim to ever more territory, the Creeks lost much of their hunting land, and thus most of their access to deerskins. What is more, those lands were still being filled by lots of hungry people, both free and enslaved. As a result, many Creeks began rearing large populations of pigs. Rather than adopting the pigs into their diet, however, the Creeks treated them as a commodity for trade with their Old World neighbors.[7] Adopting this practice changed how the Creeks settled, and they began spacing their settlements to accommodate their pigs.[8]

The situation was also fluid in sparsely settled Florida, which became an organized territory of the United States in 1822.[9] The Native Americans who inhabited the peninsula claimed a variety of animals, including horses and pigs, and they were not alone. A large number of self-emancipated African Americans (runaway slaves) also made their home in Florida. These former slaves claimed all manner of livestock, including horses and pigs, and they were often culturally absorbed into Seminole communities.[10] Everything changed in the 1830s, when the US government intensified efforts to expunge Florida's remaining Seminoles, whatever their race.[11] In 1840, Secretary of War (and South Carolina planter) Joel Poinsett procured thirty-three specially bred and specially trained Cuban bloodhounds to assist soliders in the field (Figure 4). One reporter who chanced to glimpse the animals wrote that he had never before seen such "ferocious beasts."[12] Several members of Congress cried foul, but Florida's territorial governor defended the use of dogs. Controversy persisted for years thereafter. When the general who led the Florida campaign, Zachary Taylor, ran for President in 1848, opponents reminded voters about his "heartless" use of militarized dogs.[13]

While the US Army used dogs against Native Americans in the region, a great many planters employed dogs in the subjugation of slaves. In fact, dogs became largely synonymous with the enforcement of slavery during the antebellum period.[14] Massachusetts clergyman Horace Moulton lived in Savannah during the 1830s and he confirmed that "fear of the dogs restrains multitudes from running off."[15] (Figure 5).[16] New York

Figure 4 Secretary of War honors dogs who served in Second Seminole War, 1840

The US Army famously used purebred bloodhounds to hunt down the Indians of Florida during the Second Seminole War, but not everyone approved of the policy. In this cartoon, the Secretary of War, Joel Poinsett of South Carolina, honors a troop of uniformed dogs with a flag featuring an Indian's head in a dog's jaws. Francis Preston Blair, a journalist who supported the use of wartime dogs, is on his knees pointing to a map of Florida.

**Source** – Henry Robinson, "The Secretary of War Presenting a Stand of Colours to the 1st Regiment of Republican Bloodhounds," 1840. Library of Congress. Reproduction number: LC-USZ62-91404

abolitionist George Carleton wrote that slave-hunting dogs were so numerous that they were a "domestic institution" in the South.[17] Fellow abolitionist Thomas Wentworth Higginson called dogs "the detective officers of slavery's police."[18] When artist Frederick Law Olmsted visited the South during the 1850s, he described the region's use of slave-hunting dogs in great detail. For example, he reported that runaway slaves had once found refuge in the Great Dismal Swamp in southeastern Virginia, but that enterprising bounty hunters with specially trained dogs had recently rooted many of them out.[19] Olmsted also learned that slave-hunting dogs were in high demand across the coastal regions of the

Nascent domestication initiatives and their effects on ferality 55

Figure 5  Dog chasing a runaway slave, 1855 (detail)
Dogs were synonymous with the enforcement of slavery in the late-antebellum American South. This image of a dog chasing a slave appeared as part of the frontispiece for Frederick Douglass's autobiography in 1855.
**Source** – Frederick Douglass, *My Bondage and My Freedom* (New York: Miller, Orton and Mulligan, 1855).

South, and that one breeder in North Carolina had recently sold a pack of ten hounds for more than $1,500.[20]

While slaves rightfully feared some dogs, they could, in other contexts, befriend the animals. In fact, records show that enslaved people in the South continued to own dogs during the antebellum period, just as they had during the colonial period. For example, a great many enslaved men utilized dogs when hunting game.[21] According to some historians, hunting alongside dogs provided these slaves with a modicum of dignity, and a temporary reprieve from the brutal drudgery of everyday life. As John Campbell writes, "the possession of dogs empowered slaves, making it easier for them to endure, if not weaken, the myriad assaults of slavery."[22] Some slaves even reared and sold dogs to their white neighbors for commercial gain.[23] That being said, the relationship between slaves and dogs was not merely utilitarian. Numerous sources confirm that slaves (like all people) were capable of forming deep friendships with their dogs. The life of Charles Ball provides an illustrative example. Born a slave in eastern Maryland around 1780, Ball was sold to a South Carolina planter in the early 1800s. Wrenched apart from his wife and children, Ball formed an abiding friendship with one of the dogs he met in South Carolina. He even took the dog with him when he was forced to follow new owners to Georgia. Yet when Ball later resolved to run away, he knew that he could not take the dog along. This realization caused him great anguish, and he recalled his former companion with surprising tenderness when he published his autobiography more than twenty years later.[24]

Slaves owned dogs throughout the South, but not everyone was comfortable with the arrangement. As early as 1792, George Washington insisted that African Americans should not be allowed to own dogs. "It is not for any good purpose Negros [sic] raise or keep dogs but to aid them in their night robberies," the recently re-elected President of the United States wrote.[25] Others farther south voiced similar complaints. Throughout the 1810s, numerous communities in South Carolina pressed the state legislature to pass laws prohibiting slaves from owning dogs.[26] Complaints about slaves owning dogs grew even louder in the late antebellum period. In 1858, Senator James Henry Hammond of South Carolina spoke for many of his fellow planters when he declared that "a man should not let his negroes have dogs."[27] The following year, the South Carolina legislature passed a law that taxed slave owners for every slave-owned dog on their property.[28]

Blaming slaves for the number of dogs was fairly absurd, especially since nearly *everyone* in the South retained dogs for assistance and companionship.[29] According to one observer, a veritable "dog mania" pervaded the South during the antebellum period.[30] In 1827, one planter in the South Carolina foothills observed that "a traveler in passing through this country cannot but be struck with surprise at our taste for dogs." The author confirmed that every single house in the area contained at least two "curs," and that many houses contained a half-dozen.[31] In 1832, a Savannah resident estimated that every lot in town contained three dogs, and that some lots contained as many as fifteen.[32] In 1848, a Georgia planter wrote a letter to the *Southern Cultivator* (Augusta) confirming that dogs were more or less ubiquitous in the South. "Every man in Carolina, and I suppose in Georgia also, keeps one or more dogs," he wrote, adding that "very poor men" who could not feed their families nevertheless kept four or five "useless dogs" around their domicile.[33] A year later, the same journal published another letter confirming that "there is not a farm in Georgia that has not its plantation dog."[34] In 1858, a Virginia planter wrote a letter to *Southern Planter* (Richmond) confirming that "every farmer keeps his dog."[35] Some estimated that there were more than a million dogs in southeastern North America by the 1860s.[36]

No matter who owned the animals, most southern dogs were genetically undifferentiated. There were almost no canine breeding initiatives in the South, and, as a result, very few dogs were the result of deliberately planned and prearranged couplings.[37] There were important exceptions, though. During the early national period, the wealthiest people in the South grew ever more discerning about the type of canine company they were willing to keep.[38] In 1785, George Washington wrote his friend in France, Marquis de Lafayette, about the possibility of procuring renowned French hounds. Lafayette later sent the requested hounds to Washington under the care of young John Quincy Adams, who just happened to be visiting Europe at the time.[39] A few years later, Washington wrote Lafayette with yet another request. This time, he sought the "true Irish wolf dogs," whom he hoped to employ against local wolves. Washington received a response from Irish politician Edward Newenham, who informed the future President (the inaugural inauguration was still a year away) that Lafayette had relayed the request, but that there were no more Irish wolfhounds left in Ireland.[40] This failure notwithstanding, there were many different "breeds" of dogs in Washington's kennel at Mount Vernon, including French hounds, Dalmatians, terriers, sheepdogs, spaniels, and a Newfoundland.[41]

Others tried to shape the gene pool in different ways. In 1811, a Virginia planter named Peter Minor drafted a proposal encouraging planters to promote the commonwealth's fledgling sheep industry by killing all unaccompanied dogs. "Why should we not endeavor to diminish a race of animals, which to make the best of them, are a nuisance, but when considered in a state of madness are certainly as great a curse as can visit us?" he once asked incredulously.[42] When Minor shared his proposal with his neighbor, Thomas Jefferson, the former President responded enthusiastically. "I participate in all your hostility to dogs, and would readily join in any plan for exterminating the whole race," Jefferson wrote to Minor in 1811, two years removed from his own eight-year stint as President. "But as total extirpation cannot be hoped for," he added, "let it be partial."[43] Later that year, seventy-four planters from Albemarle County petitioned the General Assembly of Virginia to do something about the scourge of feral dogs. They explained that many farmers had recently imported valuable sheep from Spain, and that packs of free-roaming dogs had subsequently killed those sheep. Their petition's stated goal was simple: "lessen the number of dogs." Rather than endorsing an extermination campaign, however, the planters suggested taxing the animals. "Such a law would reduce the present number of dogs (by) one half," the petitioners predicted. In 1814, the Virginia legislature agreed to levy a $1 tax on all dogs above six months in age.[44]

Some influenced the gene pool in more direct ways. As early as 1781, Savannah newspapers had complained about the large number of "useless dogs" roaming the streets of the town. "I flatter myself that the inhabitants will, from a principle of self-preservation, kill every dog they meet in the street," one Savannah resident remarked.[45] In 1819, officials in Charleston reminded citizens that any dog going "at large" without a human companion would be collected and destroyed without further notice.[46] In a letter that appeared in *Farmer's Register* (Richmond) and *Southern Agriculturalist* (Charleston), one anonymous contributor encouraged his fellow planters to shoot unaccompanied dogs without any more hesitation than they would shoot a mad wolf.[47] In 1852, someone using the pseudonym "Homespun" wrote a letter to the *Southern Cultivator* complaining that the South was "infested with a number of worthless, mongrel brutes, unworthy of the appellation of dogs."[48] Four years later, another farmer suggested that southerners kill 90 percent of the dogs in their midst.[49]

Some people objected to wholesale eradication, instead favoring more selective eradication. When officials in Charleston proposed exterminating untaxed dogs in 1859, a citizen using the pseudonym "Vindicator" wrote to the local newspaper in protest. "I pray you mercy, gentlemen, toward the poor helpless dogs," Vindicator pleaded. Tellingly, the author hastened to assure readers that the only dogs worth protecting were "dogs of good blood and high breeding."[50] In similar fashion, William Morton of Augusta, who expressed contempt for "the vile dog race," exempted his two favorite breeds: "the noble Newfoundland and the pointer."[51] In any event, sentimental appeals like these generally fell on deaf ears. In 1860, city officials in Macon announced that they were initiating a "war of extermination against the canine portion of our population." Although the Civil War was still several more months away, the scene in Macon unwittingly anticipated some of the horrors that would soon follow. "As the bodies of the slain were scattered over the city, we have not been able to collect any information in regard to the number of killed and wounded, but suppose the loss of dogs has been considerable as there was no quarter shown, nor prisoners taken," the *Macon Telegraph* reported.[52]

By comparison, the region's horses were subject to wildly different selective pressures. Although untold thousands of feral horses had roamed the southern wilds during the early colonial period, that was no longer the case by the early national period. As the region's human populations continued to grow, there was less room for large herbivorous equids to graze and forage. As a result, nearly all of the horses who lived in the region during the antebellum period were kept in stables or enclosed within pastures.[53] Some of these animals were subject to intense anthropogenic selection. In particular, there are numerous accounts of southern planters importing "blood horses" in an attempt to cultivate ever faster racehorses.[54] These examples notwithstanding, most horses who lived in the antebellum South did not belong to any specifically cultivated "breed."[55]

Despite their well-established affinity for horse racing, southerners did not claim the nation's largest number of horses. When officials conducted the first agricultural census in 1840, results showed that New York not only contained the most people (around 2,500,000), but also the most horses (around 475,000). The same census revealed that, New York notwithstanding, most of the nation's horses had migrated west of the Appalachian Mountains.[56] Ohio (430,527), Kentucky (395,853), and Tennessee (341,409) contained the second, third, and fifth most horses, respectively.[57] Virginia contained more than 326,000 horses in 1840, but North Carolina (166,608), South Carolina (129,921), Georgia (157,540), and Florida (12,043) each contained far fewer. Meanwhile, southern horses were not only fewer in number, but also less refined. Farmers in the Northeast and the Midwest placed much greater emphasis on "improving" animal husbandry, and their efforts produced several famous equine cultivars during the antebellum period, including Morgan horses (developed in Vermont) and Conestogas (developed in Pennsylvania), among others.[58]

The absence of widespread breeding initiatives helps explain the South's comparatively low number of horses, but there were other contributing factors. Throughout the antebellum period, for example, a growing number of southerners began to employ horse/donkey hybrids known as *mules*. Simply put, mules would not exist if humans did not deliberately intervene and compel a mare (female horse) and a jack (male donkey) to procreate. On these grounds, countless writers have compared the mule to a human artifact.[59] As one nineteenth-century writer explained, "the mule is not a natural animal, but only an invention of man."[60] There have been many hypotheses advanced to explain the preference for mules among farmers in the South, but the most intuitive reason is also the most likely. Mules combined the size and strength of horses with the durability and resilience of donkeys, and were thus perfectly suited to the South's challenging climes.[61]

Southerners obtained mules by different means. George Washington famously imported a pair of jacks from Spain for the purpose of manufacturing his own mules at Mount Vernon. The first jack, a gift from the King of Spain, was appropriately named Royal Gift. Washington advertised the animal's procreative prowess in newspapers throughout the South, and Royal Gift earned Washington more than $600 in stud fees (Figure 6). One year later, two jennets (female donkeys) and another jack arrived at Washington's estate, gifts from his old friend, Lafayette.[62] It was not just Founding Fathers who favored mules, though. Less famous farmers also sought the hybrid creatures. Their mules were not royal gifts in name or fact, and most were instead first imported from breeding farms on Cuba. In 1783, a German physician named Johann David Schoepf reported seeing mules for the first time when he was traveling in eastern Virginia. He learned that the animals had recently grown more popular, and that they could survive on "scant attention and bad feed" while dragging heavy wagons loaded with tobacco for miles over the coastal flatlands.[63] When Olmsted later passed through Richmond during his own tour of the South, he too inquired why mules were so popular, and he too learned that southerners prized the animals' durability above all.[64] Not everyone agreed, though, and many vilified the animals as products of "miscegenation."[65]

Despite the mule's popularity east of the Appalachian Mountains, the highest concentrations were actually located just west of the Appalachians, in Kentucky and Tennessee. This was because those less-populated states could provide both the grass and the grain to sustain mule-breeding operations.[66] It was unlikely that a mule born in Kentucky or Tennessee would die there, however, because each year many of them were marched eastward through the Appalachians. Cumulative statistics are difficult to come by, but even snapshot testimonies bear witness to the large volume of equids trafficked southeastward. In 1831, an estimated 1,200 horses and mules from Kentucky and Tennessee passed through Greenville, South Carolina, in a single day.[67] That same year, another 6,000 equids (4,000 horses and

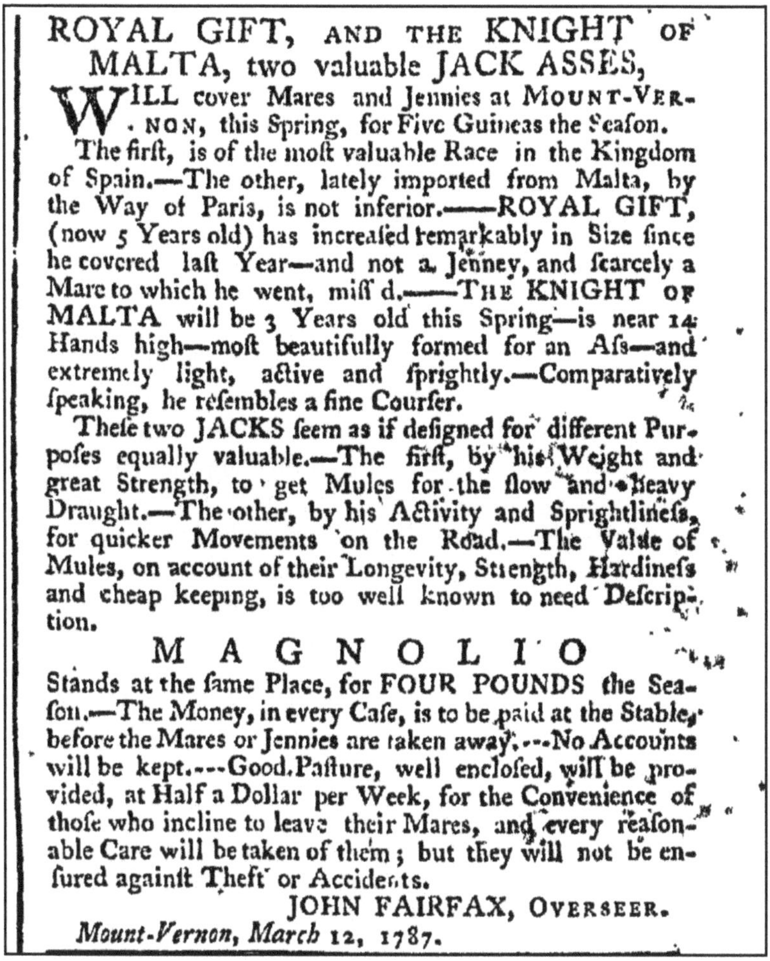

> ROYAL GIFT, AND THE KNIGHT OF
> MALTA, two valuable JACK ASSES,
> WILL cover Mares and Jennies at MOUNT-VER-
> NON, this Spring, for Five Guineas the Seafon.
> The firft, is of the moft valuable Race in the Kingdom
> of Spain.—The other, lately imported from Malta, by
> the Way of Paris, is not inferior.——ROYAL GIFT,
> (now 5 Years old) has increafed remarkably in Size fince
> he covered laft Year—and not a Jenney, and fcarcely a
> Mare to which he went, miff'd.——THE KNIGHT OF
> MALTA will be 3 Years old this Spring—is near 14
> Hands high—moft beautifully formed for an Afs—and
> extremely light, active and fprightly.—Comparatively
> fpeaking, he refembles a fine Courfer.
> Thefe two JACKS feem as if defigned for different Pur-
> pofes equally valuable.—The firft, by his Weight and
> great Strength, to get Mules for the flow and heavy
> Draught.—The other, by his Activity and Sprightlinefs,
> for quicker Movements on the Road.—The Value of
> Mules, on account of their Longevity, Strength, Hardinefs
> and cheap keeping, is too well known to need Defcrip-
> tion.
> MAGNOLIO
> Stands at the fame Place, for FOUR POUNDS the Sea-
> fon.—The Money, in every Cafe, is to be paid at the Stable,
> before the Mares or Jennies are taken away.---No Accounts
> will be kept.---Good Pafture, well enclofed, will be pro-
> vided, at Half a Dollar per Week, for the Convenience of
> thofe who incline to leave their Mares, and every reafon-
> able Care will be taken of them; but they will not be en-
> fured againft Theft or Accidents.
> JOHN FAIRFAX, OVERSEER.
> Mount-Vernon, March 12, 1787.

Figure 6 George Washington's advertisement for equine stud services, 1787

George Washington was one of the first people in the South to manufacture mules. After several well-placed inquiries, he obtained a mammoth jack (male donkey) named "Royal Gift" from the King of Spain in 1785. The following year, he obtained a sprightly jack named "Knight of Malta" from the Marquis de Lafayette. Almost immediately thereafter, the future President began advertising the animals' procreative prowess in newspapers across the South.

Source – "Donkeys," *Digital Encyclopedia of George Washington*, Mount Vernon.

URL – www.mountvernon.org/digital-encyclopedia/article/donkeys/

2,000 mules) left Kentucky and trekked east through the Cumberland Gap, bound for southeastern markets.[68] More than 3,000 mules followed the same route through the Cumberland Gap in 1838, a 50-percent increase in just seven years.[69] When British journalist James Silk Buckingham visited the South that same year, he reported that as many as 10,000 horses and mules trekked into the region from northwestern markets every year.[70] In 1843, the state-sponsored agricultural survey of South Carolina acknowledged that: "Kentucky supplies most of our horses and mules."[71]

This widespread preference for mules was not without evolutionary consequences. Most obviously, hundreds of thousands of mares throughout the United States foaled mules instead of horses. Siring horses might have helped propagate their kind, but mules were invariably sterile. Meanwhile, the mule-manufacturing industry appropriated the best horses, and the racing circuit suffered accordingly. By the 1850s, sporting enthusiasts complained that "racing is in collapse and good horses have been somewhat scarce, for pedigreed mares are now prostituted to jacks while stallions are slenderly patronized."[72] Finally, it is worth mentioning that mules almost *never* went feral. How could they? They were manufactured and purchased precisely so that they could work. That is why historian Ann Norton Greene has likened their propagation to "factory production."[73] Or, as Robert Byron Lamb once wrote, "the only reason for a mule's existence is to labor."[74]

Curiously, as the nation wrested greater control over its domestic horses, many people began to romanticize the feral ones. In 1832, a contributor to *American Turf Register* advocated: "crossing our bred horse with the wild or prairie horse."[75] The author was referring to feral mustangs of the American West, who by this point were already legendary. Two years later, John Skinner drafted an open letter to General Charles Gratiot, Chief Engineer in the US Army Corp of Engineers, on the importance of procuring the "best stallions" from the expansive prairies. Skinner acknowledged that ferality often produced "inferior" animals, but he insisted that this was not the case with the feral mustangs. Because "might is right" on the open range, he reasoned that "the enjoyments of the harem and the 'delightful task' of procreation is yielded *exclusively* to the most *active spirited* and *powerful of the herd!*"[76] A decade later, an anonymous author employing suspiciously Skinner-esque rhetoric practically snorted when repeating the claim that ferality brings out the best virtues in stallions. "As with man, 'tis liberty alone that gives to life its luster and perfume," the author wrote. "Both man and horse will degenerate in character and value when in their government there is provided no test for their capacity, no stimulus to virtue, no reward for their ambition, *nor restraint upon its vicious indulgence!"*[77]

Despite the aforementioned demand for equine power, feral horses continued to roam certain parts of the South throughout the antebellum period. Florida remained sparsely populated, and, as a result, feral horses continued to flourish during the first two decades of the nineteenth century. "The woods abound with troops of wild horses which traverse the whole peninsula," newspapers reported as late as 1821. "They are easily taken and rendered tractable by the Indians, who bring them to the European establishments, and exchange them for such weapons as they want. Their value is so trifling, that a good saddle may be exchanged for twenty."[78] Just fifteen years later, that was no longer the case. French naturalist Francis de La Porte, Comte de Castelnau, visited the peninsula in 1837 and 1838, while the Seminole Wars were still raging, and he reported that "there were formerly wild horses on Florida prairies," but that the Indians had recently massacred almost all of these animals.[79]

After this "massacre" in Florida, feral horses were even rarer in the South. Newspapers abounded with stories about the novelty of feral horses on the pampas of Argentina and the plains of the American West, indicating that they would have also been conspicuous in the South. By 1835, feral horses were already so rare that the *Farmer's Register* could print the following: "Perhaps our Virginians are not aware that their ancestors amused themselves catching wild horses. They hunted them in the uplands."[80] This indicates that feral horses had been absent from the southern range for several generations, and that any horses who remained had been pushed to the periphery. In the summer of 1845, one visitor described seeing droves of feral horses in the remote bluffs of western South Carolina. "It was a striking and imposing spectacle," he later recounted. "One moment they stood, snuffing intelligence from earth and air, then they dashed head-long out of sight, leaving behind them the crash of branches, and the faint echo of their flying feet."[81]

As previously mentioned, horses may have already reached Chincoteague and Assateague Islands as early as 1670, but, by the time they again attracted notice, people had long since forgotten how the animals got there. The islands themselves were being referred to as "terra incognita ... scarcely known to exist by most persons west of the Chesapeake."[82] The earliest published reference to the island ponies appeared in *American Turf Register* in September 1829. The article noted that "small horses, known by the name of Beach Ponies," roamed various islands along the southeastern seaboard, from Delaware to Georgia, but otherwise did not elaborate.[83] Thompson Holmes, a physician who lived on the Delmarva Peninsula near Chincoteague Bay, described the animals at length in 1835.[84] *Farmer's Register* had recently published a query asking its readers if these fabled ponies actually existed, and Holmes responded with a letter to the editor answering in the affirmative. He explained that the horses technically belonged to a group of men who lived on the peninsula, but that the horses received no attention during the year. He added that the horses on nearby Assateague Island were rounded up during an annual "penning" event that attracted thousands of visitors one weekend every summer. Holmes explained that men on domestic horses would chase the feral horses in an angular pen made of logs and driftwood, and then try to subdue the animals. The whole frenzied scene was marked by "unrivalled noise, uproar and excitement, which few can imagine who had not witnessed it, and none can adequately describe."[85]

Even more interesting, however, are the myths that sought to explain how these horses first arrived on Chincoteague and Assateague. After all, they quite obviously owed their presence on the seaboard islands to human intervention one way or another, and yet no one could remember when or how they got there. By the time they were first mentioned in the historical record, they were already being romanticized as relics of a bygone era. In Holmes's letter to the editors of the *Farmer's Register*, he wrote that the horses had been decreasing in number and stature for the previous three decades.[86] When a correspondent for the *Hartford Daily Courant* inquired about the pennings in 1848, he too learned that, "like everything poetical connected with the habits of our people, this custom is rapidly becoming obsolete, and will soon be remembered merely as an idle and romantic tale."[87] As we shall see, this tendency toward romanticization would eventually influence the fate of the islands' feral horses for years to come.

Chincoteague and Assateague were not the only islands with feral horses. In the 1850s, Virginia agriculturalist Edmund Ruffin, the editor of the aforementioned *Farmer's Register*, confirmed that hundreds of feral horses also roamed along the Outer Banks of North Carolina. He noted that the islanders in the Outer Banks also held an annual penning festival, in which they too wrested horses from ferality back into domesticity. There were few people in eastern North Carolina who had not attended the annual penning at least once. As for the horses, Ruffin noted that they were exceedingly shy, and that they would flee from people if approached. He believed the present stock had "suffered deterioration by the long continued breeding without change of blood." He suggested importing new males from elsewhere on the Outer Banks, but hastened against importing purebred horses. "It would be the reverse of improvement to introduce horses of more noble race, and less fitted to endure the great hardships of this locality. Such horses, or any raised in other localities, if turned loose here, would scarcely live through either the plague of blood-sucking insects of the first summer, or the severe privations of the first winter," he wrote.[88]

Meanwhile, pigs were subject to unique selective pressures, quite distinct from those shaping dogs and horses. During the early colonial period, European settlers in the Americas had relied on the stock they brought from their homelands, which is to say they relied on European pigs. That was about to change. The reader will recall that pigs were independently domesticated in several different places, including China, and thousands of years of intensive (and relatively isolated) domestication had endowed the Chinese pig with unique morphological and physiological properties. As one nineteenth-century farmer explained, Chinese pigs showed "a remarkable tendency to fatten."[89] This lattermost quality, the propensity to fatten quickly at a young age, was the Chinese pig's most important quality, and British farmers began importing Chinese pigs in large numbers around 1700.[90] Consequently, when farmers in the Northeast and Midwest began cultivating distinctly "American" breeds like the Duroc Jersey, the Chester White, and the Poland China during the early Republic, they were actually working with genes that hailed from China, by way of Great Britain.[91]

No one is entirely certain when the first Chinese pig arrived in the Americas, although, remarkably, one of the earliest sightings was actually in the South. In 1783, the aforementioned Schoepf noticed a Chinese pig when he passed through St. Augustine. The animal was distinguished by "short feet, hanging, dragging belly, and softer bristles." Upon inquiry, Schoepf learned that the pig had arrived on a ship from the distant East Indies.[92] The fate of this animal is unknown, though he or she was presumably eaten. In any event, the animal did not help establish a southern swine industry, and appears not to have contributed to the regional porcine gene pool, either. The earliest references to Chinese hogs in the Midwest appear around 1805, though they may have arrived earlier, since the exact origins of particular breeds were often obscure. For example, the Poland-China apparently emerged around Butler, Warren, and Hamilton Counties (Ohio) in the 1830s, though more precise information remains elusive.[93]

Things were drastically different in the South. While farmers in the Northeast and Midwest had enclosed their commons and "improved" their pigs, farmers in the South showed no interest in selectively breeding their pigs, to say nothing of fencing them in. Southern pigs still had the legal right to roam wherever they pleased during the antebellum period, and a greater majority of them still exercised this right. According to historian Sam B. Hilliard, these animals were "an admixture of various strains and had degenerated to the point where the characteristics of the parent European breeds were indistinguishable."[94] Without any anthropogenic oversight, the pigs were instead shaped by natural selection. They grew leaner, meaner, and exceedingly quick on their feet.[95]

Southerners did not consider their method of rearing swine remarkable, and, as a result, visitors from outside the region offered the most detailed descriptions of southern pigs during the antebellum period. When Buckingham toured southern Appalachia in the late 1830s, he observed that "hogs were everywhere abundant," but that they were scarcely recognizable.[96] His disdain was unequivocal, and his description is worth quoting at length:

> These are among the ugliest of their species, with long thin heads, long legs, arched backs, large lapping ears, lank bodies, and long thin tails, and they are among the filthiest of the filthy. I had never before thought there could be such difference in pigs; but I may now say, that the hog of England is as much superior in beauty of form and cleanliness of habit to the hog of America, as the Bucephalus of Alexander was to the Rosinante of Don Quixote; as superior, in short, as animals of the same race can be to each other.[97]

Olmsted made similar observations when he toured the South during the 1850s. As he saw it, southern pigs were "long, lank, bony, snake-headed, hairy, wild beasts" who dashed across the landscape in large, intimidating packs.[98] In fact, feral pigs were so numerous and so tenacious that Olmsted took them into account when planning his route through eastern Virginia. He was not averse to adventure, he wrote, but he had no intention of ever encountering a feral pig at night.[99]

While most of the region's pigs were unfenced, that does not mean that they were also unclaimed. Many people marked pigs with distinguishing earmarks and then set them loose to roam the commons, thereby continuing a practice that first began during the colonial era. Even so, proving and enforcing ownership always proved difficult, especially since the pigs readily mixed and procreated with the region's other marked and unmarked pigs. To be sure, some animals were kept closer at hand. For example, many homesteads in the South contained adjacent enclosures for livestock. Some plantations enclosed pigs on a year-round basis, though smaller farms more frequently enclosed them on a seasonal basis.[100]

With the obvious exception of cotton, few other organisms are more closely associated with the antebellum South than pigs.[101] As Hilliard once observed, "nowhere in the nation at any time were swine as important in a region's dietary resources as they were in the South during the three or four decades prior to the Civil War."[102] Indeed, pigs were a staple in nearly every southerner's diet. When Buckingham visited the South in the 1830s, he noted that pigs were "the universal food of all classes."[103] Emily Burke offered similar testimony when she moved to Georgia in the early 1840s. "Bacon, instead of bread, seems be their staff of life," she wrote.[104] Olmsted reported the same when he visited the region in the 1850s, insisting that he had once been offered "seven different preparations of swine's flesh" at a single meal.[105] To satiate their voracious appetite, every community in the region held annual "hog killings." During these arduous but highly anticipated events, families and neighbors helped one another slaughter, salt, and store enough pigs to last through the winter.[106]

Meanwhile, a great many slaves also owned pigs.[107] Olmsted confirmed as much when he passed through the rice plantations of the South Carolina Lowcountry in the 1850s (Figure 7). "The negroes' swine are allowed to run in the woods, each owner having his own distinguished by a peculiar mark," he observed.[108] Even though slaves had owned pigs since the colonial period, many planters objected that "allowing" slaves to raise pigs, or to hold property of any kind, provided the slaves with leverage, and was thus unacceptable. "Every measure that may lessen the dependence of a slave on his master ought to be opposed, as tending toward dangerous consequences," one group of South Carolina planters explained in 1816.[109] Throughout the 1830s, citizens in eastern North Carolina complained that "allowing" African American slaves to possess property had rendered the local slaves well-nigh "uncontrollable."[110] Despite these loud complaints, European Americans of every social station, from "poor whites" to plantation owners, engaged African Americans in the underground economy.[111]

68  *Feral Animals in the American South*

Figure 7 Enslaved people and their animals in the South Carolina Lowcountry, 1856
Frederick Law Olmsted sketched this scene while passing through the vast rice plantations of the South Carolina Lowcountry. He reported that many enslaved people owned pigs, whom they allowed to roam free, unenclosed.
**Source** – Frederick Law Olmsted, *A Journey in the Seaboard Slave States: with Remarks on their Economy* (New York: Dix and Edwards, 1856), 423.

Needless to say, however, slavery was not a get-rich scheme. Slaves toiled every day of their lives under a system that extracted their labor through coercion, brute force, and, ultimately, the imminent threat of death.[112] Given these deplorable conditions, it is not altogether surprising that untold thousands of slaves tried to escape bondage during the antebellum period. Many of these self-emancipated slaves sought refuge in the vast expanses of land where other humans would not settle. The largest such "maroon" population was located in the Great Dismal

Swamp of southeastern Virginia, which Kirby described as "perfect terra incognita."[113] As early as 1784, there were reports that runaway slaves had survived in the swamplands for more than thirty years, subsisting on the seemingly endless supply of feral pigs.[114] Similar reports from elsewhere in Virginia and North Carolina confirmed that runaway slaves often subsisted on feral pigs. When bounty hunters discovered an abandoned maroon community in the South Carolina Lowcountry, they noted that the runaways had even constructed a pig pen in the swamp.[115]

Suggestively, the first people in the South to start agitating about the number of feral pigs were people who owned large numbers of slaves in densely settled eastern Virginia.[116] John Taylor of Caroline County, who owned more than a hundred slaves, was among the first people to suggest closing the southern range. He published his ideas in a treatise titled *Arator* (1814). His assertion that horses and mules "ought to inhabit a lot having a stable and stream, and to be excluded wholly from grazing," was not controversial. As previously stated, most southerners understood that equids were too large and too valuable to leave on the commons. Where he differed from most of his contemporaries, however, was his disdain for feral pigs. "Few animals do us more real mischief, and suffer more unmerited reproach, than these [pigs]," Taylor complained. Furthermore, he continued, allowing pigs to fend for themselves meant that half of the South's pigs were destined to perish in the vast southern wilds, while the remainder was subject to theft. As a result, farmers utilized less than half of their potential pig crop during any given year. "If hogs were prohibited from going at large," he wrote, "these pains and penalties would be vastly diminished." Taylor recognized that closing the southern range would drastically reshape southern society and culture, but he insisted that it was the more economical choice. "By prohibiting hogs from running at large, we should be compelled to recur to a new mode, which might possibly be more adequate to our wants," he wrote.[117]

Although the southern range remained squarely intact throughout his life, John Taylor found an enthusiastic acolyte in the aforementioned Edmund Ruffin, who launched the first serious attack on the southern range in the 1830s. To promote his agenda, Ruffin began publishing an agricultural journal, *Farmer's Register*, in 1833. Within five years, the journal reached approximately 4,000 wealthy planters, mostly in Virginia. It is impossible to know just how many of the journal's articles can be attributed to Ruffin since they were often published anonymously or under pseudonyms, but historians credit him with drafting many of them, especially those attacking the open range.[118] As early as

1834, Ruffin complained that the current fence laws requiring people to enclose plants instead of animals were a "most intolerable nuisance." First and foremost, there were few fences that could resist attacks from feral pigs.[119] In addition, however, Ruffin also objected on a fundamental level. He pointed out that people cannot just trespass anywhere they want, but that the law protected animal trespasses.[120] Whether they were authored by Ruffin or not, articles on the southern range in general (and feral pigs in particular) filled the pages of the *Farmer's Register*. One contributor named "Jeremiah" acknowledged that closing the range might not be possible, but nevertheless advocated for a law that would "compel everyone to keep his hogs in an enclosure."[121]

Ruffin's *Farmer's Register* was not the only agricultural journal in the antebellum South to debate the merits of feral pigs on the open range.[122] One unnamed planter from "Lower Virginia" wrote a letter to *Southern Planter* insisting that the fence laws favored the rights of "half-starved hogs" over the rights of people.[123] Here again, the anonymous contributor acknowledged that popular sentiment was against closing the commons, but that something should at least be done about the feral pigs.[124] The scourge of pigs even prompted these antebellum planters to do something shocking: appeal to the Yankee's superior husbandry. In 1834, one planter from South Carolina informed his fellow southerners that states in the Northeast did not allow pigs to roam at large, and that "the law there requires each man to keep up his own stock."[125] Others were more incredulous: "Is the common mode of rearing hogs in the Southern States an economical one?" one planter asked. "Could you not procure from some experienced person at the North the best mode of constructing and managing a piggery?"[126]

In 1858, Ruffin used his real name when he submitted a committee report to the Farmers' Assembly on the law of enclosures in the commonwealth of Virginia. The committee, which consisted of Ruffin and two other planters, implored legislators to do something about the state's fence law, or, at the very least, its feral pigs. Their very first recommendation was that "the law be so modified as to prohibit the running at large of hogs in Eastern Virginia. It is now almost universally conceded in this portion of our State, that hogs can be raised and fattened more economically within our enclosures, and in sties, than ranging at large on the barren commons."[127] Lawmakers did not close the range, though they did afford Ruffin and his neighboring planters in Prince William County a modest victory when they passed a law allowing for "ring fences," which encircled *groups* of farms and within which stock law would reign.[128]

Despite the pigs' apparently ubiquitous distribution south of the Potomac, this region did not contain the nation's highest number of pigs. In 1840, the first agricultural census revealed that most of the nation's pigs had already moved west of the Appalachian Mountains. The largest number of pigs lived in Tennessee (2,926,607), Kentucky (2,310,583), and Ohio (2,099,746). By comparison, Virginia (1,992,155), North Carolina (1,649,716), South Carolina (878,532), Georgia (1,457,755), and Florida (92,680) contained far fewer pigs. Meanwhile, the "razorbacks" who roamed the southern range stood in sharp contrast to the corn-fed behemoths on trans-Appalachian farms. By 1860, the average pig in the Chicago stockyards weighed approximately 228 pounds, whereas the average pig on the southern range weighed well below 140 pounds.[129] Pigs born west of the Appalachians did not always die there. Instead, untold millions were funneled eastward through winding mountain passes so they could be sold in southeastern markets, where the demand was always much higher. Significantly, many of these pigs were fed to the South's increasing number of enslaved African Americans.

In 1790, the first federal census counted more than 544,000 enslaved African Americans living south of the Potomac River.[130] By 1860, there were more than 1,740,000 slaves living south of the Potomac, almost all of whom were born in North America.[131] Planters required a lot of pigs to feed their enslaved labor force, and pork rations for slaves had become more or less standardized by the 1840s. Allowing for numerous variables (work load, season, availability of other foodstuffs, etc.), historians generally accept 3½ pounds of pork per week as the "standard" slave ration.[132] It follows, therefore, that each slave consumed approximately 150 pounds of pork per year, on average, and that all the slaves in the five southeastern states consumed more than 260,000,000 pounds of pork every year by 1860. Southern planters were famously devoted to cotton and other cash crops during the antebellum period, and they showed little interest in devoting their land to free-ranging pigs. Rather than culling wiry razorbacks from the surrounding landscape, many planters found it more cost- and time-effective to simply purchase pigs as needed, and the intra-regional pig trade flourished as a result. Even so, demand exceeded supply, prompting many planters to import pigs from beyond the Appalachians.[133] Since these trade patterns developed long before railroads, pigs were driven into the South on the hoof. Routes were well worn and well known, and generally followed the routes traced out by equid traders.[134]

As was the case with equine importations, cumulative numbers are difficult to come by. Once again, however, the evidence suggests that an *enormous* number of pigs were driven into the South.[135] In the 1830s and 1840s, approximately 60,000 pigs left Kentucky and passed eastward through the Cumberland Gap every single year.[136] Meanwhile, an even larger number of pigs annually marched eastward through the valley of the French Broad River.[137] Edmund Cody Burnett, who was born near the French Broad River, explained what it was like to grow up in that area as a boy. From his fixed vantage point, he watched a seemingly neverending herd of pigs funneled from farrowing operations in Tennessee to hungry bellies in North Carolina, South Carolina, and Georgia. "It did seem as if all the world were hogs and all the hogs of the world had been gathered there, destined for the Carolina slaughter pens and the cotton-growers," he wrote. There were times when pigs were backed up for miles and miles, stretching over the horizon, awaiting their turn at the ferry. There are no official statistics, but Burnett guessed that 100,000 pigs passed by every year.[138] These traffic jams grew smaller during the 1850s, as more and more pigs were shipped east of the mountains on rails (Figure 8).[139]

This trade deficit left the region vulnerable when relations with the northern states turned sour, though historians have often overlooked this fact. Even today, it is not uncommon for historians to claim that the South boasted a larger number of pigs than the North on the eve of the Civil War.[140] To arrive at this claim, these authors include pigs from two states (Missouri and Kentucky) that sanctioned slavery but never actually seceded. When one adds their pigs to the northern tally, the scales tilt in the opposite direction. The Union states could call upon more than 17,300,000 pigs, while the Confederacy contained just below 15,600,000.[141] Furthermore, the South's pigs were, according to one historian, "pitiable animals." Georgia contained more pigs than any other Confederate state east of the Appalachian Mountains (about 2,000,000), but they were, according to one historian, "long-headed, long-legged, fleet-footed 'piney woods rooters' used to depending on providence for food."[142]

Many southerners recognized their precarious position with regards to pigs. "We are caught unprepared. We have much to sell, nobody to buy, and little to eat and wear," one panicked planter wrote to the *Southern Cultivator* in January, 1861. "Never were a people suddenly overtaken by an emergency in a worse state of preparation in other respects, than the people of the South at this moment."[143] Others

Nascent domestication initiatives and their effects on ferality    73

Figure 8  Pigs being driven to railroad cars, 1857
Although this illustration of pig drovers depicted an age-old practice, things were starting to change by the late 1850s. A reporter for *Harper's Monthly* encountered these pigs being driven from Abingdon, Virginia, to Bristol, where they would be loaded onto railcars and shipped to markets farther south.
**Source** – "A Winter in the South," *Harper's New Monthly Magazine*, vol. XV, no. LXXXIX (October 1857), 595.

voiced similar concerns. "The droves of hogs transported by our Railroads are enormous in number," one southern planter pointed out in March, 1861. "The drain arising from this cause to the Cotton States is a serious evil. We speak of dependence upon the North ... we are more hurtfully dependent upon the West."[144] Planters took to the pages of agricultural journals to implore one another to start raising more pork immediately. "We have so long bought Cincinnati and Western pork," one writer noted, "that it has almost become second nature with us to look away from ourselves for everything except the raw article of Cotton and Corn bread."[145] Although the war had scarcely begun in July, 1861, planters were already starting to point their fingers at one another in blame.[146] It was not just citizens, however. From the very beginning, Confederate officials recognized that the demand for pig flesh would present a major problem. Lucius Bellinger Northrop, head of the Confederate Subsistence Bureau, acknowledged that the "real evil is ahead. There are not hogs in the Confederacy sufficient for the Army and the large force of plantation Negroes."[147]

The mobilization of Union forces meant that untold thousands of northern soldiers, all of whom hailed from places where the commons had long since closed, were able to observe the South's unique method of free-range pig husbandry for the first time. In November, 1862, First Lieutenant Charles Francis Adams (who was the great-grandson of the nation's second president, John Adams) took note of the "long-nosed, lank Virginia swine" in his diary.[148] While stationed in eastern Virginia in May, 1863, Union general George Henry Gordon reported that soldiers indiscriminately killed feral pigs in the area, "justifying themselves by affirming that the pigs ran loose in the woods, well knowing that the whole State of Virginia was one vast pigpen."[149] Gordon's observation that Virginia was "one vast pigpen" could have applied equally well to the entire South. In September, 1863, Henry Harrison Eby of Illinois was captured following the Union defeat at Chickamauga. While he was being transported by rail to Belle Island Prison near Richmond, Virginia, he noticed the feral pigs of South Carolina. "The hogs we saw in the woods were so thin that two of them were required to make a shadow," he remarked.[150] Another soldier stationed in western Virginia declared that the South's feral pigs were the "longest, lankiest, boniest animals in creation."[151] Although the pigs sometimes provided food and/or target practice, they also provided nightmares. After some skirmishes, feral pigs roamed the battlefield, gorging on dead soldiers.[152]

Southern agriculturalists continued calls to increase domestic pork production. In late 1862, one farmer acknowledged that "the pork business of the great North West has taken millions of dollars from the South and now that that source of supply has been cut off, we think that all that portion of the Confederacy denominated the 'Bread States' ought at once to exert their energies to produce those things on which the cotton planters are to live."[153] Helping matters but little, most of Tennessee (as well as its pigs) had fallen to the Union by the end of 1862. As Confederate territory continued to shrink, farmers could do little more than lament their lack of foresight. "Of course it is impossible now to give out the old rations of meat to our Negroes, but the improvidence of our people heretofore, in raising home supplies, is at the bottom of the whole matter," one farmer lamented in December, 1863. "Had we exercised proper foresight, industry and thrift in years past, we could have supplied our army and our people abundantly with meat, and all the other necessaries of life."[154]

The problem was not just that the Confederacy had lost access to its largest source of pigs. After all, there were still millions of the animals in the South, but the region's relatively pathetic infrastructure left no way to move them toward the battlefront other than marching them hundreds and sometimes thousands of miles.[155] Moreover, what few railroads the Confederacy possessed were ill-equipped for wartime traffic.[156] When hungry soldiers in Virginia demanded more pork from the Confederacy's hinterlands, Tennessee was ill-equipped to assist. By comparison, the Union was in a much better position to utilize its huge pool of pigs. Between 1850 and 1860, railroads had laid thousands of miles of track between the Union's urban centers in the Northeast and its fresh-meat market in the Midwest. Railroad connections between New York City and Chicago were completed in 1851, while Philadelphia and Baltimore were connected to the Ohio River by 1852 and 1853, respectively. By 1855, most farmers in the Northeast and the Midwest were relatively close to railroad lines.[157] What is more, many meatpackers had begun experimenting with refrigeration techniques by 1857, thereby dramatically increasing output.[158] This was a matter of no small consequence. In Chicago, the great railroad hub of the Midwest, pork production increased six-fold during the first three years of the war. By the end of 1862, Chicago was the largest pork-packing center in the United States.[159] Once again, the differences between North and South were stark. "Although hardly immune from hunger, northern soldiers received more food per person than any other army in the history of warfare," historian Ted Steinberg writes.[160]

Despite the importance of provender, it was not pigs but rather horses who would determine the fate of the war. Many participants in the war felt that the Confederacy possessed certain advantages when it came to horses. After all, most of the horses in the Union army were draft animals used to pull carriages, but there were not enough miles of improved roadways to warrant a large number of carriages in the South. Instead, people in the region rode their horses in the saddle, an arrangement that prepared them well for mounted cavalry duty. Reflecting on his experiences in 1866, Pennsylvania native Samuel Ringwalt admitted that the draft breeds of the northern states had proved "sadly deficient" for mounted warfare, and he bemoaned "the lamentable inefficiency of our cavalry in the early days of the war."[161] Historians have generally agreed, ascribing an early advantage to the Confederate cavalry.[162]

Here again, however, from a strictly numerical standpoint, the Union possessed an undeniable advantage. When we include those slave states that never seceded among the Union tally, the northern states had access to a far greater number of horses. Whereas the Union contained more than 3,170,000 horses on the eve of war, the Confederacy contained just 1,740,000. It is true that the Confederate states contained an absolute numerical advantage when it came to mules. In 1860, the Confederacy contained more than 800,000 mules, over 70 percent of the nation's total, yet this surplus was hardly enough to overcome their relative dearth of horses. When one combines the number of all horses, asses, and mules, the Union had access to more than 4,640,000 equids, while the Confederacy still contained fewer than 2,570,000.[163]

During the first year of the war, the Confederate quartermaster could still conscript horses from Virginia, relatively close to the front.[164] It did not take either side long to understand that horses conveyed extraordinary advantages. Lieutenants on both sides directed their soldiers to aim for their enemies' horses, which contributed to an enormous number of equine casualties.[165] And since horses do not just disappear when they die, their carcasses became a seemingly ubiquitous symbol of war.[166] While horses in the war's eastern theater were being killed, those in trans-Appalachian parts of the Confederacy were being captured. Most of Tennessee, which contained the second-most horses in the Confederacy after Texas, had fallen into Union hands by the summer of 1862.[167]

Not surprisingly, Confederate officials and their supporters began to worry about how long their horse supply would hold out.[168] Quartermasters from both the eastern and western theaters found themselves competing for the Confederacy's limited supply of horses.[169] Officials struggled to produce viable solutions. In 1862, the Confederate Secretary of State, Judah P. Benjamin, suggested rounding up horses in Texas to send to Virginia, but nothing ever came of the idea.[170] Making matters worse, horses required enormous amounts of forage, but the lands surrounding the massive Army of Northern Virginia were already exhausted by the end of 1862.[171] In the spring of 1863, General Robert E. Lee drafted a letter in which he openly worried about the availability of Confederate horses. "The destruction of horses in the army is so great that I fear it will be impossible to supply our wants," he wrote, adding that: "there are not enough in the country."[172] Indeed, when Lee finally surrendered to General Ulysses S. Grant at Appomattox Courthouse in 1865, it was, in no small part, because he lacked the horses necessary to outpace the Union army.

According to the most recent estimates, more than 750,000 soldiers died during the American Civil War.[173] That is a staggering number of deaths. Though they have attracted less notice than their human compatriots, millions of domestic animals also lost their lives during the war. The number of horses in the briefly Confederate states decreased by 32 percent during the war, while the number of mules decreased by 37 percent. Losses were particularly acute in the southeastern seaboard

Figure 9 Horse statue outside the Virginia Historical Society in Richmond, Virginia

Designed by Tessa Pullan, this statue commemorates equine sacrifices during the Civil War. According to our best estimates, the southern states lost more than a third of their horses during the war. Appropriately, the animal is portrayed as gaunt, tired, and war-weary. The same statue (though smaller in size) is located at the National Sporting Library in Middleton, Virginia, and the US Cavalry Museum in Ft. Riley, Kansas.

**Source** – Photograph by Lee Brauer Photography, Richmond, Virginia.

states. Virginia's horse population decreased by 40 percent during the war, while Georgia's horse population decreased by 45 percent. The destruction of southern swine was no less dramatic. The Confederate states had laid claim to more than 15,000,000 on the eve of war. After five years of war, they now contained fewer than 8,900,000, a 43 percent decrease across the entire region.[174] All told, the Confederate states lost somewhere between one-third and one-half of their livestock. Among these animals, none played a more important role in the outcome of the war than horses. In recognition of this fact, people in the South have erected several prominent monuments to the Confederate warhorse (Figure 9).[175]

*Chapter Summary*

People were unevenly distributed across the South during the earliest years of the republic. Virginia and South Carolina were becoming more densely settled by people of European and African ancestry, but large swaths of the South remained peopled by Native Americans, whose population densities were generally much lower. As the number of people living in the region continued to rise, animal populations began to cleave along previously ill-defined lines. These bifurcations meant different things for the various populations involved. For example, some dogs found their gene pools more tightly controlled than ever before. People became very particular about what kinds of dogs would be allowed to procreate with other kinds of dogs. These genetic cultivars, also known as "breeds," were employed in a variety of tasks, but, all too often, people of European descent used these purebred dogs as weapons of terror against peoples of American or African descent. Meanwhile, a much larger share of the South's canine genome were free-roaming, genetically undifferentiated dogs known as "curs," "mongrels," or "mutts." Because neither their range nor their reproduction was monitored by people, these dogs proliferated greatly, and they wreaked havoc on farmers' sheep.

The selective parameters acting on horses were significantly different, of course, and so the South's equine population cleaved along its own unique lines. As the human population rose, there was necessarily less room for large herbivorous animals such as horses to roam. By the time the first agricultural census was taken in 1840, demographic pressures had pushed most horses west of the Appalachian Mountains, where they found ample forage and range. As a result, farmers east of the mountains had to import their equids from the west. By the late antebellum period, many farmers in the South were beginning to favor "mules," a deliberately manufactured horse/donkey hybrid who was designed to work but who was incapable of reproducing. All the while, the last shrinking bands of feral horses persisted in the region's most remote recesses. Some populations carved out an existence on the sparsely settled barrier islands dotting the South Atlantic seaboard, while others took refuge in Florida's most inaccessible reaches.

Finally, pig populations also began to cleave during this era, though a drastically different set of selective parameters shaped their fortunes. While horses had been deemed too big and too valuable to remain at large, the southern range was otherwise still wide open. As a result, the vast majority of southeastern hogs remained free-roaming throughout the antebellum period. A few enterprising agriculturalists in the most densely settled parts of eastern Virginia began to attack the open range in general, and feral hogs in particular, but these individuals were vastly outnumbered and easily shouted down. Despite the vast army of hogs roaming the southern wilds, however, there were never enough to fulfill the South's insatiable appetite for pork. The shortage was particularly acute in those parts of the South most ruthlessly devoted to cotton. As a result, southeastern planters annually imported millions of swine from west of the Appalachian Mountains. These pigs were originally driven into the region on the hoof, though railroads had begun taking over some of the traffic by the late antebellum period.

Few events have transformed the history of the South more than the Civil War, which wreaked havoc on humans and nonhumans alike. Pigs were in higher demand than ever. Rambling armies required lots of energy, which required lots of protein, which required lots of pigs. Despite historiographical opinions to the contrary, the Confederate states actually contained fewer pigs than the Union states on the eve of war. Meanwhile, excursions into the South provided many Union soldiers with their first glimpse of feral hogs. Of course, no animal proved more important to the outcome of the war than the horse. Whether carrying supplies or people, horses played an integral role in determining how the war was executed. Equine attrition was remarkably high on both sides, but the Union army had access to far more horses than the Confederacy, and could replenish its front lines with less effort. Five years of war devastated the South's populations of domestic animals. According to our best estimates, the region's supply of pigs and horses diminished by more than a third during the war.

# 5

## ANTHROPOGENIC IMPROVEMENT AND ASSAULTS ON FERALITY

### DIVERGENT FATES IN THE INDUSTRIALIZING SOUTH

> *Americans are forever searching for love in forms it never takes, in places it can never be. It must have something to do with the vanished frontier.*
>
> ~ Kurt Vonnegut

Although the nation's swine population had been utterly devastated during the Civil War, it rebounded in relatively short order. By 1880, there were more than 47,000,000 pigs in the United States, almost double the number from 1870. The nation's swine population grew even larger over the next two decades, and there were more than 62,000,000 pigs in the United States by 1900. The increase would have been even higher, but the southern states acted as a drag. Thirty-five years after the Confederacy collapsed, they had still not recovered their prewar complement of pigs.[1] In fact, Iowa and Illinois contained three times as many pigs (approximately 15,600,000) as all five southeastern states combined (approximately 4,700,000).

There were several reasons for the South's paltry supply of pigs. Unbalanced wartime devastation provides a partial answer, but the pigs' legendary fecundity would have been more than capable of overcoming that dearth in relatively short order. A more important factor was the nature of southern agriculture at the time. While plantation slavery had died with the war, the planters' ethos had not. As a result, many farmers in the region remained staunchly committed to an economy based on cash crops. The new (but not unfamiliar) practice of sharecropping allowed some planters to retain debilitating control over their recently emancipated neighbors, and it allowed cotton to remain king in many places throughout the remainder of the nineteenth century. Farmers were still not keen on committing vast stretches of potential cropland to rangeland, yet there was still huge demand for pigs in the South, where more pork was eaten than all other meats combined.² Rather than raise their own animals, many continued importing pigs in huge numbers, thereby expanding a practice that they had initiated during the antebellum period.

The pigs imported by southerners were usually born in the Midwest, where farmers had always taken far more interest in "improving" their livestock. Because pigs who fattened quickly increased productivity, and therefore padded the bottom line, Midwestern meatpackers relied more and more on pigs of Asian origin, particularly the aforementioned Poland-China, whose propensity toward childhood obesity rendered the animal extremely valuable.³ Indeed, the pigs were far more valuable than corn, and so it made sense to feed as much corn to the pigs as was biologically possible. As it turns out, that was *a lot*. No two states in the union contained more corn than Iowa and Illinois, and the USDA estimated that more than 30 percent of the nation's annual corn crop ended up being fed to pigs every single year (see maps in Figure 10).⁴ It was therefore only slightly tongue-in-cheek when one observer rhetorically asked, "What is a hog, but fifteen or twenty bushels of corn on four legs?"⁵ By the late nineteenth century, the Poland-China was the nation's most popular, most numerous and most widespread type of pig.⁶

Unlike in the antebellum period, however, these pigs were no longer marched into the South on the hoof. Instead, they arrived on the nation's rapidly expanding network of railroads, and they achieved remarkable infiltration.⁷ By 1890, 90 percent of people in the South lived in a county with at least one railroad station.⁸ Some openly wondered whether the South still deserved its reputation as the land of "hog and hominy" since most of its pigs and most of its corn originated elsewhere. To be sure, the USDA tried to promote better swine farming in the southern states, especially through youth programs like the 4H, but they first had to disabuse southern farmers of their traditional ideas about animal husbandry (Figure 11).⁹ Perry Van Ewing, author of *Southern Pork Production*, certainly agreed, adding that the South lagged far behind other regions when it came to animal cultivation.¹⁰ He estimated that 0.001 percent of pigs in the region were "purebred." Worse yet, he wrote, whenever some enterprising farmer actually *did* procure purebred swine, they invariably encouraged the animals to roam at large, where they "indiscriminately mixed with other blood."¹¹

The increased reliance on Midwestern hogs necessarily decreased reliance on the South's free-ranging swine. Although the commons had

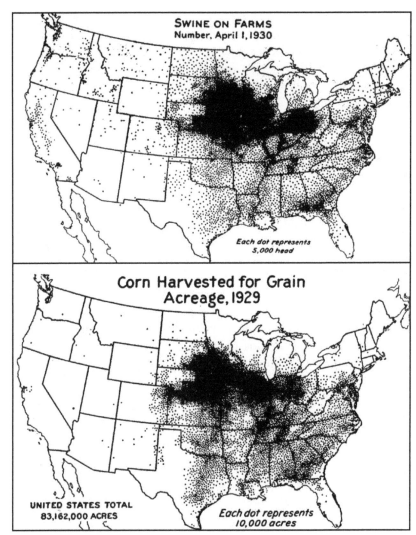

Figure 10 Maps showing distribution of pigs and corn, 1929–1930
As these maps clearly demonstrate, the biogeographical distribution of pigs was closely tied to the biogeographical distribution of corn.
**Source** – Earl Shaw, "Swine Production in the Corn Belt of the United States," *Economic Geography*, vol. 12, no. 4 (October 1936), 360.

allowed pigs to run wild throughout the antebellum period, conditions changed after the war. As mentioned in the previous chapter, the first minor changes to fence laws in the South occurred just before the war, when Virginia legislators allowed wealthy planters to form "ring fences" in King William and Prince George Counties.[12] The next substantial

Figure 11 Cyrus Martin and feral pig, Ossabaw Island
The South's free-ranging pigs were significantly smaller than their corn-fed cousins in the Midwest. They did not eat nearly so well, certainly not as much, and they constantly burned through calories as they rooted about in search of mast. The feral pigs of Ossabaw Island had already endured ferality for several centuries by the time island caretaker Cyrus Martin bagged this characteristically small specimen in the early twentieth century.
**Source** – Catalog no. 1326-62-06-01, Georgia Historical Society, Savannah, Georgia.

change actually occurred *during* the war, in 1862, when the General Assembly of Virginia delegated authority over the commons to county administrators. By the end of the year, most of the counties in eastern Virginia had closed the open range.[13] Others were not far behind. In the late 1860s, while the South was still reeling from war, citizens in Augusta asked their fellow Georgians to consider revising the fence law. Not surprisingly, free-ranging pigs lay at the heart of their complaints. One citizen explained that revising the law would "kill off" excess pigs and "improve" those who remained under human control. Similar conditions prevailed near the capital of South Carolina, where planters likewise struggled against an onslaught of "wandering hogs." Nevertheless, support for the open range remained strong throughout the 1860s. "Any fool can see that if we adopt this new plan the poor at least, and that means four-fifths of our rural population must give up utterly every vestige, every hoof, of domestic animals," one citizen explained incredulously, unable to conceptualize a scenario in which rural people lived without access to animals.[14]

Throughout the 1870s, similar referenda met with similar results. In 1872, North Carolina, South Carolina, and Georgia all passed compromise legislation that would allow each county within those states to decide the fate of the commons by popular votes.[15] Despite vigorous attempts by planters, a voting bloc of poor whites and recently enfranchised poor blacks consistently thwarted attempts to close the open range.[16] In 1873, six counties in North Carolina held votes on revisions to the fence law, and all six rejected any such changes.[17] Interestingly enough, the federal government made no effort to close the southeastern range during Reconstruction. Instead, the tipping point occurred when Reconstruction ended and federal troops withdrew. This change left southeastern legislatures more squarely within the hands of wealthy planters, who were quick to disfranchise those who would oppose their grand design. The plan was most effective in South Carolina, where planters exercised the most control. In 1879, William M. Porcher drafted a petition to revise the fence law around Charleston. He reported that feral animals sometimes wandered twenty to thirty miles away from their homestead, and he implored the legislature to change the existing fence laws to prevent roving pigs from destroying the state's crops.[18]

The open range in Georgia collapsed in piecemeal fashion throughout the 1880s and beyond. William M. Browne of the Georgia Department of Agriculture first endorsed revisions to the fence law in 1880, but Georgia's planters could not coerce the region's poor, stock-owning people to close the range. Therefore, legislators delegated the authority to districts *within* counties. This naturally generated even more debate among citizens, and, more often than not, the question hinged upon pigs. In 1882, one citizen wrote that wherever the open range was closed, "razor-back hogs ... have given place to comely, easily-kept, productive ... hogs at once beautiful and profitable." By 1890, thirty-seven counties had closed the open range entirely, another thirty-six had closed the range in parts of the county, and the remaining sixty-one counties had not yet put the question to a vote. Slowly but surely, the southeastern range was collapsing.[19]

One's opinion on the rationality of existing fence laws depended almost entirely on where one lived. People who lived in the South's growing number of townships generally opposed the open range, while people who lived in the county (but not in town) fought just as stringently to keep the fence law, and thus the open range, intact. In addition to the obvious urban/rural component to the fencing controversy, there was also a North/South component.[20] This was most obvious in locations with resort country clubs, where northerners tended to congregate. For example, when citizens of Richmond County, Georgia, voted on the fence law, those living in the county seat of Augusta voted to close the range by a 3-to-1 margin, while those living outside the town's limits voted to keep the range open by an 8-to-1 margin. Events in North Carolina provide an even more illustrative example. Displaced northerners who lived in the town of Pinehurst had recently forced revisions to the local fence law, and the Moore County citizens who still relied on free-ranging animals were none too pleased. When police began dutifully impounding animals, the county folk decided they could take no more. Several marched into town and sawed down the pen where their animals were impounded. When police rebuilt the pen, the county folk returned, madder than ever. According to reports, about fifty people marched into town carrying "guns, axes, and staves, cut the pen down and made bitter speeches and threats."[21]

Whether they liked it or not, rural southerners were fighting a losing battle. In those parts of the region where pigs still roamed free, calls grew louder to close the commons (Figure 12). In 1908, B. F. Keith, customs agent in Wilmington joined those who advocated for a new fence law. "It is a well-known fact that ... the piney woods rooter in those counties which they are permitted to run at large do not bring in revenue enough to their owners to pay for keeping up their fences around their farms." Making matters worse, he continued, feral hogs destroyed the region's long-leaf pine forests. His interests were not ecological, but rather economic. Enclosing pigs would allow pine forests to regrow, which could then yield turpentine and rosin. While the range collapsed across most of the South, however, it remained open across all of Florida (save for

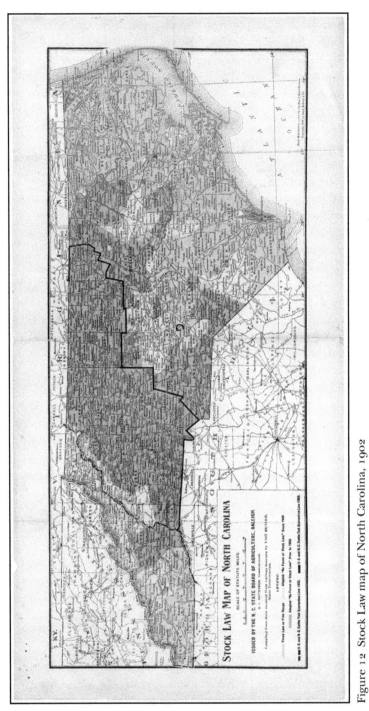

Figure 12 Stock Law map of North Carolina, 1902

Few topics inspired more debate among people in the South than the fate of the open range. As this map of North Carolina demonstrates, one's opinion on existing fence laws largely depended on where one lived. People in the more densely populated Piedmont counties (the shaded counties on this map) had voted to close the commons by 1900, while those in the less populated coastal and mountain counties (the lighter counties) fought to keep it open.

**Source** – "Stock Law Map of North Carolina," North Carolina, State Board of Agriculture, 1908. University of North Carolina Libraries.

a few scattered towns). Occasionally, some northerner would move to Florida and start making noise about the absurdity of the open range, but these individuals were shouted down rather quickly. Those who wanted to change the fence law and close the commons were dismissed as "irrational and unreasonable fanatics."[22]

Scholars have only recently begun to consider how closing the commons affected the evolutionary trajectories of animals. It was a known maxim that when fence laws were revised, the number of pigs in that area dropped dramatically. In 1905, reports from Heathsville (Northumberland County, Virginia) revealed that hogs had more or less disappeared ever since the county voted to close the range. Similar reports surfaced in North Carolina, where the razorback increasingly gave way to "better" breeds. According to reports, feral pigs had almost "disappeared and in their place is the Berkshire, Poland China, Cheshire, or other hog that responds more readily to proper raising." In 1910, *Forest and Stream* confirmed that feral pigs were fast disappearing from the Tar Heel state. In 1916, R. L. Shields, Chief of Animal Husbandry at Clemson Agricultural College bid the feral pig adieu. "The ungainly, unprofitable hog known as the razor-back is fast giving way in South Carolina to a better type," he wrote.[23] In 1922, a reporter for *Meat and Livestock Digest* boasted that the South's feral pigs were not long for this world. "Somebody ought to have a genuine Florida razorback mounted before they become extinct," the journalist reported, adding that "the porker around which so many yarns have been spun soon will be a candidate for the museum."[24]

While it is true that feral pigs disappeared from some parts of the South, reports of their extinction were exaggerated. They were never completely eradicated, persisting in the region's rugged mountain terrain and coastal riverine shallows. They invariably caught the attention of visiting northerners, who generally had no prior experience with free-ranging pigs. "The Florida hog seems to be of less domestic importance than his fellows in the north," one understated contributor to *Forest and Stream* reported in 1876, adding that "he is little cared for, and being left to his own devices assumes an independent and self-reliant character." Other visitors to Florida were surprised to discover that "large herds of hogs are raised, and roam about the forests half wild." One reported that the number of free-ranging pigs sometimes grew so cumbersome that hunting parties would annually set forth to winnow the herds. "The razorback hog as known in Florida is seldom subjected to favorable conditions for development. He is ranged, starved, ill-treated, forced to shirk for himself and seldom actually subjected to favorable conditions, and rarely systematically, intelligently or adequately fed or fattened for market before being slaughtered."[25] Florida was not unique in this regard. After all, the commons remained open in other parts of the South, like swamps and Sea Islands, where population densities remained especially low. Nor were razorbacks confined to coastal lowlands. Many also roamed the Appalachians. When Charles Sloan Reid described these creatures for *National Stockman and Farmer*, explaining that the "mast hog seems as thoroughly at home among the rocks as does the mountain goat, and stands preeminent in his hideousness."[26]

Ferality endowed pigs with unique morphological traits. Tobe Hodge tried to explain the "razorback" nickname to his Northeastern and Midwestern readers. "This species of hog takes its name from its likeness to a razor with the thin edge up, the sharpness of its vertebral column," he wrote in 1887 (Figure 13). Four years later, J. M. Murphy, a writer for *Outing Magazine*, reported that the southern pig's legs were "long, lean and sinewy, its hocks short, its body attenuated to the verge of the ridiculous, its snout prolonged and tapering, its skull low and elongated, its neck scrawny and its back arched in the centre and sloping gradually toward the flanks." Unlike hogs born in the Midwest, who lived their entire lives in sties, southern hogs were constantly foraging for sustenance, and, as a result, they never grew fat. Meanwhile, S. M. Shepard described the southeastern pig as "long nosed, long eared, long necked, long legged, slab-sided, small hammed, coarse haired, large bristled, gaunt, restless, hard feeder, and an impudent 'cuss.'" Not to be outdone, Reid wrote that the razorback's "spinal column assumes a slight bow shape, and along the whole length of it stand up stiff, sharp bristles." In 1901, Leonidas Hubbard, an editor at *Outing* magazine, offered similar descriptions. "What a change came over the pig as it dropped away from the ways of its cultured ancestors and became once more a savage!" he marveled.[27]

While northerners obviously noticed the South's feral hogs and delighted in describing these curious creatures, they did not originally regard them as potential hunting trophies. This is significant, because, as historians are well aware, hunting was very popular among the northeastern elite in the late nineteenth century. These sportsmen began by hunting the areas closest to them, the undeveloped areas of the Northeast, where there were no feral pigs. When northeastern hunting grounds were denuded of their game, sportsmen turned their attention

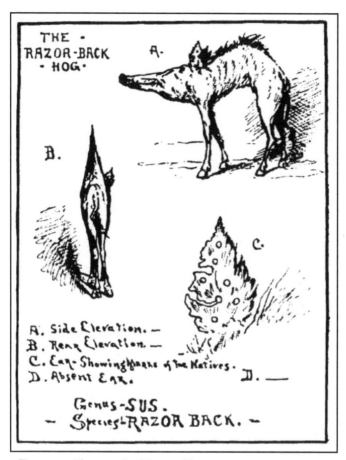

Figure 13 The razorback hog, 1887
Explaining the feral pig's nickname, Tobe Hodge wrote that "this species of hog takes its name from its likeness to a razor with the thin edge up, the sharpness of its vertebral column."
**Source** – Tobe Hodge, "Razor-Backs," *The American Magazine*, vol. vii, no. 2 (December 1887), 255.

southward. One old sportsman lamented in *Field and Stream* that "the old haunts of the North have become drained." By 1891, Walter F. Mickle wrote in *Forest and Stream* that tropical and sparsely populated "South Florida offers the finest sporting of any section of the U.S. that I have seen."[28]

Northerners were able to access these southern hunting grounds for the same reason that Asian-derived hogs were now able to reach the region: because railroads had recently begun penetrating the most remote recesses of the South. Railroads advertised access to the South's fabled hunting grounds. Outfits like the Seaboard Air Line Railway promised prospective hunters that excellent game abounded in the southeastern states.[29] Meanwhile, Midwestern newspapers reported that no region promised more bountiful grounds than the South: "The 'New South,' the south of industrial regeneration, is becoming a sportsman's heaven." Like so many others who visited the South, these sporting tourists soon discovered that the southern fauna included feral pigs. In 1897, the *New York Times* informed its readers that a party of New Yorkers had recently been treed by feral pigs near the head waters of the St. John's River. Five years later, *Forest and Stream* shared a story about a group of northern hunters who chanced to encounter (and kill) a feral pig while in Florida. The hunters, for whom hog killing was already something from a bygone era, were "much pleased at the prospect of reviving 'hog killing' days ... when laughing negroes, the hoe cake and the cracklings, made the day a festal on the old plantation home."[30]

Around the turn of the century, newspapers in the Northeast and Midwest informed their readers that a new kind of "game" (often in quotation marks) awaited sportsmen who traveled to the South. "Sportsmen are not accustomed to think of wild hogs as game that may be hunted in the United States, but the fact remains that there are thousands upon thousands of the animals wandering through certain sections of the South," Hubbard wrote in the *New York Times*. "They are as wild as deer and well-nigh as formidable as the bear." Reid agreed, insisting that "the razor-back hunt is exciting sport to those who understand it." These non-southerners found it exceedingly strange that southerners did not consider feral hogs a sporting target.[31]

As word of the South's feral swine spread, recreational pig hunting grew increasingly popular in the coastal regions of Georgia and South Carolina. There were several reasons for this. First, the commons remained almost exclusively open throughout the coastal lowlands, and razorbacks thrived in the region's riverine flats. Second, and no less important, the region's rice industry had more or less disappeared by the turn of the century. Whereas the South Carolina lowcountry had committed 78,000 acres to rice production in 1879, the same region committed just 19,000 acres thirty years later. Once-thriving plantations fell into disuse and once-glorious mansions slowly fell into ruin. According to J. William Harris, coastal plantation lands were "receding into the wilderness from which they had been created." Enter the tourist sportsman. Throughout the late nineteenth and early twentieth centuries, speculators from the Northeast and the Midwest began buying up old, dilapidated plantations and turning them into exclusive hunting clubs.[32]

William Whitson related one such visit to the Santee Hunting Club in January, 1899. He and his fellow sportsmen had discovered the hunting lodge was once a magnificent mansion. Their hunting guide belonged to "a distinct species ... of old-fashioned South Carolina Negroes," who spoke a unique dialect. The guide was almost certainly Henry Snyder, the club's most famous hunting guide, whose Gullah dialect delighted northeastern visitors. Whitson's hosts had informed the tourists that "wild hogs are frequently found in these woods," but neither Whitson nor his friends believed the claim. Their hosts reiterated that "if any of you see a hog to-day you may be certain he is a wild one and shoot him." By the end of the first day, they had killed their first feral hog.[33]

A very similar pattern was repeated on the Sea Islands. Perhaps the most famous hunting club was located on Jekyll Island, Georgia. In 1886, a cohort of millionaires, including J. P. Morgan, William K. Vanderbilt, Joseph Pulitzer, and Marshall Field, purchased the island for $125,000. At the time of purchase, the island contained about 600 feral pigs. The millionaires originally wanted the animals exterminated, and so they contracted a hunter, Mr. Lamb, to accomplish the task. The pigs, however, failed to cooperate, and Lamb met with limited success. The millionaires hired another professional hunter, Captain W. R. Townsend, to exterminate the pigs, but he met with similar results. As a result, the pigs were neither eradicated nor seriously affected.[34]

Eventually, the millionaires discovered what many other tourist sportsmen had discovered: that the feral hog provided thrilling sport. In 1889, Montgomery Folsom reported that "the clubmen find no end of sport in hunting these bristly boars, with trained dogs and rifles. They frequently send fine specimens of boars head north, where they are prepared and mounted." Contemporary sources estimated that there were several thousand feral pigs on Jekyll Island, "including some old boars that would rival for size, length of tusks, and fierceness those of the Black Forest." There was a glorious, never-ending campaign to kill the feral hogs, "affording opportunities for the Nimrods of the club to show their prowess."[35]

Hubbard believed that people in the South would eventually recognize that feral pigs provided new financial possibilities. "The time is not far distant when the Southern hunters will find some way to have a sport with this sort of game which will be comparable with that furnished by the wild hog of Europe," he predicted. Sure enough, they soon recognized that feral pigs were an economic product. One Midwestern newspaper reported that southeastern citizens had begun treating feral pigs "as if they were gold," and that many hotels now offered hunting excursions. Meanwhile, newspapers advertised southeastern islands for sale with the promise that the island contained feral pigs.[36]

During the first two decades of the twentieth century, the domestic and feral pigs living in the South were also joined by their distant cousin: the wild (never-domesticated) boar. King Victor Emmanuel III of Italy sent a pair of the animals to American diplomat Lloyd C. Griscom in 1909. "I completely forgot about the King's wild boar until one day I received a telegram from a frantic shipping agent, saying that they were at the New York docks," Griscom later recalled. At a loss, he began asking his friends what to do with the animals. Following J. P. Morgan's suggestion, Griscom released the animals on Jekyll Island, "where they were allowed to mingle with the indigenous wild pigs." Three years after that, additional wild boar were also introduced into the mountains of western North Carolina. A financial advisor from Detroit named George Gordon Moore had established a game preserve on Hooper Bald. In 1912, he released thirteen wild boar into a fenced enclosure. The animals remained unmolested for about a decade, and so their numbers continued to grow. They escaped into the surrounding mountainsides during an organized hunt in the early 1920s. In short order, these wild boar began to interbreed with the untold number of feral pigs who were already inhabiting the mountains. When Leroy Stegeman analyzed the free-ranging suids who roamed the Smokies in 1938, he confirmed that the population showed evidence of significant admixture.[37]

As may well be expected, horses were subject to radically different selective pressures than pigs. It was once assumed that railroads would replace the nation's horses, but it did not take long for people to realize that the opposite was actually true. As early as 1860, the agricultural census had recognized that railroads actually *increased* demand for horses, often dramatically, and that this close interrelationship was an established principle. As historian Ann Norton Greene explains, railroads covered long distances efficiently, but horses were still essential for short-distance hauling to and from railroad depots.[38] Consequently, the nation's horse population soared following the end of the Civil War. The population more than doubled between 1870 and 1900, from roughly 7,000,000 to roughly 17,000,000. No region saw its horse population grow more than the Midwest, which benefited from the postwar expansion of railroads connecting the eastern states to the western territories. In some cases, the growing demand was met by sweeping feral mustangs off the western range.[39] The rate of increase in the southeastern states was slower, but otherwise not so dissimilar from the national trend. Between 1870 and 1900, the number of horses in the five southeastern seaboard states almost doubled, from roughly 393,000 to 672,000. But this increase was misleading. After all, it was not until 1900 that the South regained its prewar complement of horses. And whereas the five southeastern states accounted for roughly 10 percent of the nation's total in 1860, they accounted for less than 4 percent in 1900.

Feral horses had been celebrated for their novelty before the Civil War, but they were even more celebrated after the war. In 1877, the feral horses on Chincoteague garnered nationwide attention when *Scribner's* ran a feature on the island horses. Every few years thereafter, another article would appear and inform another generation of readers that, yes, feral horses persisted in pockets along the eastern seaboard (Figure 14). It was especially curious for the more industrialized northerners who lived relatively close to the eastern shore of Virginia, and yet seemingly a million miles away. "It will somewhat astonish the people who hasten across the trackless wastes of lower Broadway or wander aimlessly over the boundless steppes of Madison-Square to learn that within less than 250 miles of Battery Park great droves of wild horses roam at large the whole year through," a reporter for the *New York Times* reported in 1884.[40]

By that point, it was customary for the islanders to hold a penning event on Chincoteague one day, and on Assateague the next day. Attendance increased as the years passed by. "Of late years, Chincoteague has been visited every June by large numbers of New York and Boston people, who come down on steamers to witness the exciting spectacle of the annual roundup," one local oysterman confirmed in 1902. It is significant and instructive that no one could actually remember how these animals got to the islands. In lieu of an actual explanation, almost every single article that was written about the Chincoteague ponies credited their origin to a mythical shipwreck centuries earlier. More often than not, these reports ascribed the horses a Spanish pedigree.[41]

Meanwhile, hundreds of feral horses also persisted along the Outer Banks of North Carolina. Like their more celebrated cousins on

Figure 14  Escorting feral horses from Chincoteague back to Assateague, 1887

In 1887, *Scribner's* reporter Howard Pyle informed his readers that horses "as wild as the mustangs of Texas" inhabited the Sea Islands off the eastern shore of Virginia. He explained that locals rounded up the feral animals during a lively penning festival every summer. "Many of the ponies are taken over the narrow channel that separates Chincoteague from Assateague, to run wild upon the latter island, which is largely unclaimed land," he wrote. "We were fortunate as to witness the lively scene of the swimming of a number of ponies across this channel or inlet."

**Source** – Howard Pyle, "Chincoteague: the Island of Ponies," *Scribner's Monthly*, xiii, 6 (April, 1877), 743.

Chincoteague and Assateague, these banker ponies were allowed to roam free their entire lives, save for one weekend each summer when they were collectively rounded up and penned. But the Outer Banks are even more remote than Chincoteague, which is saying something, and so the pennings never attracted the same level of revelry. Nevertheless, those lucky enough to attend a penning on the banks around the turn of the century would not soon forget it. Locals did not use lassos to subdue the animals, but instead used their bare hands and their steel courage. Some of the horses were loaded onto small barges and taken to the mainland to be sold. According to contemporary sources, they were in "great demand in the middle part of the state," and had been known to sell for as much as $125. It was hard to beat the investment. The market for horses was stronger than ever, and the herds on the Outer Banks "cost literally nothing, breeding in droves as wild horses." The horses who were not ferried across to the mainland were branded and turned loose, left to fend for themselves until the next go-round.[42]

Another similarity that Banker ponies shared with the Chincoteague herd was that no one had any clue how they got to the islands, and so a similar origin myth emerged. In 1902, reports claimed that the horses were descended from stock brought over during the late-sixteenth century by Sir Walter Raleigh's colonists. One year later, *Forest and Stream* repeated that North Carolina's "tackies" (a colloquial name for marsh horses and Banker ponies) owed their origin to horses abandoned by Sir Walter Raleigh centuries earlier.[43]

There were also feral horses on the islands off the coast of Georgia. Recall that there were about 600 feral pigs on Jekyll Island when some of the richest people on Earth purchased the property. Well, in addition to those swine, there were also about 200 feral horses.[44] Many of these animals were gunned down as the new owners of the island decided to "wage a war of extermination on the wild horses." They did not shoot all of them, though. Instead, attempts were made to capture and sell some of the horses. Montgomery Folsom reflected on these efforts a decade later. "I saw the only horse trap, I guess, I will ever look upon," he recalled. "By hook and crook they were all caught and sold." Some of them ended up on nearby Cumberland Island. Another set of millionaires, Thomas and Lucy Carnegie, had recently purchased the island, and they allowed Jekyll's refugee horses to join the "half-wild horses" who had already roamed the island since before the Carnegies arrived.[45]

The first gasoline-powered automobiles in the United States had been imported from Europe in the late-1880s, but it was not long until Americans began building their own automobiles. As might be expected, most of the earliest automobiles were located in the Northeast and the Midwest. In 1900, the state of Pennsylvania contained 1,480 automobiles. That same year, the five southeastern states contained 220, *combined*.[46] There were several reasons for the automobile's relatively slow arrival in the South. First, the earliest automobiles required wealth, and there is no denying that most of the nation's capital was centralized in the Northeast and the Midwest during first few decades of the twentieth century. Second, automobiles required state-of-the-art paved roads, but the vast majority of southeastern roads remained in poor condition in the early part of the century.[47]

Eventually, of course, the automobile *did* find its way into the South. The first automobile arrived in Savannah in 1899, but others were not far behind.[48] Charleston hosted its first automobile show that same year, and organizers advertised the event far and wide. Advertisements promised curiosity-seekers that, in addition to the automobile, other modern wonders (including typewriters and "grapho-phones") would also be on display, but there was never any question which invention was the most dazzling. "The automobile is 'the thing' these days, and the interest in these horseless carriages is becoming a craze," the *Evening Post* promised, adding that the "automobile has not yet struck Charleston," but it was on the way.[49] "The expert opinion is that the horseless carriage has come to stay," the *Daily Enquirer* (Columbus, Ga.) confirmed.[50]

There were just over 200 automobiles in the five southeastern states at the turn of the century, but that number increased significantly in the years thereafter.[51] "In cities of the rank of Atlanta, Augusta, and Macon, the noiseless, swift movers are seen to rush by in great numbers," the *Macon Telegraph* reported in 1902. Of course, this was old news in other parts of the nation. "In the larger cities of the North they have already become so numerous as to be no longer objects to attract observation," the *Telegraph* reported.[52] In due course, the urban populations in the South grew similarly conditioned. At the turn of the century, an automobile was "a novelty on the streets of Augusta, a machine looked upon in the light of a new plaything for the rich, but of no particular interest to the average man except as a target for alleged jokes in the funny papers." By 1908, however, the town boasted more than 140 automobiles. "Gradually it has increased in importance until now the buzz wagons have become just as much a part of the local economy as horses, bicycles or any other modes of locomotion, while the garages and repair shops are on an almost equal footing with the livery stables," the *Augusta Chronicle* reported.[53]

The first automobile races in the region were held in 1903, on the wide, smooth beaches that stretched between Ormond Beach and Daytona Beach.[54] More than 3,000 people attended the race. Most of the people in attendance were wealthy northerners, men and women who had long since vacationed in sunny Florida, but the crowd also contained a "sprinkling of open-eyed and open-mouthed 'crackers' and 'pickaninnies.'"[55] Ormond and Daytona hosted additional races in the years thereafter, and, in relatively short order, the *New York Times* declared northeast Florida "a mecca for motor enthusiasts," adding that its beaches could provide "undoubtedly the greatest straightaway course for racing in the world."⁵⁶ Automobile enthusiasts in Macon hosted that city's first automobile race in 1906, while Savannah and Atlanta hosted their first automobile race in 1908 and 1909, respectively.⁵⁷ Citizens had initially supported these races, which attracted teams from around the world, but their enthusiasm quickly waned. As Randall L. Hall explains, "the increase in local ownership of automobiles created much opposition to closing the roads for days, as the races necessitated."⁵⁸

Although the South's earliest automobiles were restricted to the region's scattered towns and cities, there was little doubt that they would soon become available to rural populations as well. "As the machines are perfected more and more and grow cheaper and cheaper, they will become familiar farther and farther from the towns," one Georgia newspaper predicted.⁵⁹ As late as 1911, however, automobiles were still few and far between on southeastern farms. "The use of the automobile is confined almost exclusively at present to those seeking recreation and pleasure," one South Carolina newspaper reported, though it predicted that "next year will find the farmers – those living at a distance from the railroads – using the automobile truck for transporting their fertilizers to the farm."⁶⁰ Eventually, of course, the region's rural population began to recognize that the automobile offered unparalleled advantages. Using horsepower, farmers were limited to the surrounding ten or twenty miles. This meant that they more or less practiced subsistence farming. Automobiles promised to change all that. The machines opened up whole new opportunities for "trucking" agricultural products to consumers hundreds of miles away.⁶¹ Farmers were able to cover far more territory, in less time, at lower costs. By 1930, about 40 percent of southeastern farms contained at least one automobile.⁶²

The number of automobiles continued to rise in every single community every single year, and the implications for the horse were lost on no one. The *Charleston Post* had predicted as early as 1897, when there were only a handful of automobiles in the state, that citizens would no longer need the horse's services in the future.⁶³ By 1899, there were still no more than forty automobiles in South Carolina, but Columbia's major newspaper nevertheless saw the writing on the wall. "It is probably only a question of time when it will replace most of the horse drawn vehicles on city streets and country roads," the *State* proclaimed.⁶⁴

Some thought this development would ultimately be *good* for the horse. One citizen in Macon wrote that "raw-bone back horses might breathe a sigh of release in anticipation of being set free." He predicted that the national Society for the Prevention of Cruelty to Animals would welcome the automobile's wider diffusion, since the organization had heretofore labored so hard to ameliorate the mistreatment of horses. He encouraged his fellow citizens to respect the state's increasingly superfluous horses, and to provide them pasture if they were no longer needed. He proposed: "that when the army of worn-out hack horses is released by the advent of the automobile, the municipal government in each community should provide a pasture wherein the aged and infirm animals might die in peace and comfort." This is an opportunity to be humane, he wrote, and help the "faithful old beasts that have given their lives to help mankind."[65]

Many southerners were confessedly "bitter against the automobiles," and they fought their arrival tooth and nail. In 1903, a group of farmers in Columbus, Georgia, implored county commissioners to restrict the use of automobiles. Their most frequent complaint was that automobiles frightened horses, who were not yet accustomed to the machines. "The roads are narrow and the puffing machines frighten the horses, and a good deal of feeling has been worked up among the farmers," the *Enquirer* reported. One citizen proclaimed that "the next time he was going along a country road in his buggy or wagon and one of these automobiles tried to dash by him he was going to use a shotgun to try and prevent it – remarking that he would prefer a tragedy to a runaway."[66] Similar complaints surfaced in western North Carolina. According to the *Chronicle*, "Mecklenburg [County] stock, having never made the acquaintance of horseless carriages, are apt at the sight of a 'Red Devil' to take to the woods or fields, their intellects so completely unhinged that they take no notice of fences and ditches in their going."[67] While some rural people in the South bemoaned the arrival of the automobile, there was little one could say in consolation. "If he [the horse] does not like automobiles he has got to get used to them, and that is all there is about it," the *Augusta Chronicle* reported in 1899.[68] Six years later, the same paper doled out similar advice to the aggrieved farmers of western North Carolina. "The mules and horses of Mecklenburg will have to grow accustomed to the autos. Nothing can withstand the march of progress."[69]

Others were equally convinced that the horse, who had served alongside humanity since antiquity, could never, and would never, disappear.[70] They cited the fact that, two decades after the arrival of the automobile, the nation's horse population continued to rise. Strange though it may sound, the numbers supported their argument. At the turn of the century, approximately 16,900,000 horses lived in the United States. By 1920, that number had risen to more than 19,700,000. The South experienced a similar increase. Approximately 672,000 horses lived in the five southeastern states in 1900. Twenty years later, the number had risen to more than 700,000.[71] Milton P. Jarngain, Head of Animal Husbandry at the Georgia College of Agriculture, assured animal lovers that horses were not going anywhere. "I believe that the draft horse will always remain the most reliable and economic source of farm power," he wrote.[72] As late as 1925, Ellsworth Huntington dismissed the suggestion that automobiles would ever replace the horse's utility. "The introduction of motor transportation is sometimes supposed to have eliminated the horse as a necessity. This is not true," Huntington insisted. "The effect of motor transport in diminishing the number of horses in the United States appears almost to have reached its limit."[73]

Huntington and the others had been unduly optimistic. Beginning around 1920, the nation's horse population, which had plateaued during the previous decade, began to plummet. Though more than 19,700,000 horses roamed the nation in 1920, fewer than 13,600,000 remained just ten years later. Approximately 500,000 horses disappeared every year during the 1920s, and though some died of old age, many were sold to meatpackers so they could be processed into dog food, leather, and glue.[74] By 1940, there were barely 10,000,000 horses left in the entire nation.[75] By comparison, the number of automobiles in the United States rose from approximately 9,200,000 in 1920, to more than 32,400,000 in 1940. Meanwhile, the horse's trajectory was even more dramatic in the South. Though more than 700,000 horses had lived in the five southeastern seaboard states in 1920, fewer than 281,000 horses remained in 1940.[76] By comparison, automobiles in the South increased from around 569,000 to more than 2,472,000 during that same twenty-year span.[77]

Meanwhile, the mules' experiences were slightly different. As mentioned in the previous chapter, mules had populated the South since the earliest years of the republic, but it was only *after* the Civil War that the mule became synonymous with the region.[78] There were just 146,000 mules living in the five southeastern states in 1850, but there were more than 960,000 in 1924. In fact, mules drastically outnumbered horses in the region throughout the 1920s. As late as 1935, by which point the number of horses in the southeastern states had fallen below 376,000, there were still more than 945,000 mules roaming the region.[79] That is because the mule's closest analogue was not the automobile, but rather the tractor, and the tractor had not yet arrived in the South en masse. Once again, King Cotton was to blame. "There are few tractors in the Cotton Belt, for the acreage of cotton a farmer can handle is usually limited by the acreage he can pick by hand," USDA literature explained in 1930.[80]

Eventually, however, things began to change. Whereas the southeastern states had contained fewer than 8,900 tractors in 1920, the same five states contained more than 104,000 tractors by 1945.[81] Change was in the air and the demand for mules slowly but surely began to wane. This invariably had a profound effect on southern landscapes. In the span of a generation, garages replaced stables, and farmers who could once grow their own power and fuel became reliant on distant machine and oil companies.[82] One can hardly understate the effect this shift had on people who worked with mules. Prior to the arrival of the tractor, Pete Daniels notes, "many farmers spent more time with their mules than they did with their wives and children."[83] In the span of a generation, everything changed. "Farmers became mechanics who worked sitting down," Kirby writes, "and those who lived through the transformation never overcame the feeling of loss especially of the *company* of an animal."[84]

With so little demand for domestic equids, one might think that the demand for feral horses would disappear altogether, yet this was not the case. In fact, the invention of the automobile meant different things for the various pockets of feral horses inhabiting the South. We will begin with the feral horses on Chincoteague and Assateague. The first automobile arrived on Chincoteague Island in 1909.[85] There was no bridge connecting the island to the mainland, so the automobile had to be ferried over. The number of automobiles on the Eastern Shore of Virginia remained relatively low at the time, however, and so there was still a demand for the islands' horses. According to contemporary sources, there were "probably 600 wild ponies on the two islands" in 1911.[86] The annual pennings continued, and the reclaimed horses routinely sold for more than $100.[87] In 1913, some of the horses sold for as much as $125.[88] Those purchased were ferried back to the mainland on small barges.[89]

Soon, however, things began to change. There had only been 2,040 automobiles in the entire commonwealth of Virginia when the first one visited Chincoteague in 1909. A mere eleven years later, there were more than 115,000 automobiles registered in the commonwealth.[90] By November 1922, a causeway connecting Chincoteague to the mainland was completed. Two weeks later, approximately 200 automobiles crossed the bridge every single day.[91] The dramatic proliferation of automobiles led to a dramatic drop in the demand for horses. In 1910, the average cost of a horse in Virginia was $142.50. By the summer of 1923, the cost had fallen to $85.00.[92]

Meanwhile, a major fire tore through downtown Chincoteague in 1924, the second conflagration on the island in just four years. In response to the second fire, the town sanctioned a volunteer Fire Department. To fund the new Fire Department, the town would host an annual carnival, timed to coincide with the annual pony penning. The firemen, who did not actually own any of the horses, were charged with rounding up the ponies on Assateague and ushering them to Chincoteague. Since there were far more horses than boats, the firemen just directed the horses to swim across the channel. The horses were rushed ashore, herded through the streets of Chincoteague, and enclosed in a holding pen near the center of town. The first pony penning under the auspices of the firemen attracted more than 15,000 people, and was roundly considered a huge success.[93]

In 1925, the Chincoteague horses were sold at fixed prices, $75 for males and $90 for females. These prices were already well below the prices of the 1910s. Just four years later, however, the price had dropped even further. "Now, the demand for them is more limited, and buyers at $25 to $50 apiece are none too plentiful," an observer from the *Baltimore Sun* reported in 1929.[94] Three years later, stock owners reverted to an auction, essentially allowing buyers to name their price.[95] A report the following year confirmed that "average prices have dwindled steadily in recent years." Nevertheless, attendance at the pony pennings continued to grow every year. By the 1930s, the pony penning annually attracted 25,000 people. And so the ritualized festival continued unabated, not because there was any real demand for these horses, but because there was a demand for the communal experience of the penning, a practice that had defined life in these parts for the previous 300 years. It is perhaps significant that the origin myth surrounding these horses intensified during the automobile era. In fact, there were several myths. The most popular held that the horses were descended from a European shipwreck.[96] In 1938, Marshall Andrews reported that "the generally accepted version of the ponies' origin is that a few were marooned on the islands by the wreck of a Spanish ship centuries ago."[97] Jack Woodson, a noted Virginia artist, painted a dramatic mural in the Hotel Russell on Chincoteague dramatizing the horses' noble (if unfounded) pedigree.

Meanwhile, a different fate awaited the feral horses of North Carolina. The herds on the Outer Banks had always been significantly larger than those on Chincoteague and Assateague. In 1902, Fred Olds, a North Carolina native, estimated that there were approximately 1,200 feral ponies on Shackleford Banks alone (the southernmost section of the Outer Banks).[98] The following year, *Forest and Stream* estimated that there were around 3,500 feral horses along the entire Outer Banks.[99] As demand for the animals on the mainland dropped, the number living on the Banks increased. By 1925, Olds estimated that there were now approximately 3,000 feral horses living on Shackleford Banks.[100] One year later, Melville Chater, a writer for *National Geographic*, observed immense droves of feral horses in the Outer Banks. He estimated that there were between 5,000 and 6,000 horses on the islands at the time.[101] Chater observed as "a chosen dozen were auctioned off at about $6 a head." He learned that the same horses would have sold for more than $100 apiece a few years earlier. When he inquired about the low prices, an auctioneer explained, "too much gasoline about nowadays!"[102]

By the 1930s, the number of horses living on the Outer Banks had grown to such an extent that authorities decided to destroy them. Beginning in 1935, an extermination campaign was initiated, and hundreds (perhaps thousands) of horses were shot down. According to reports, the extermination campaign was executed by county deputies, who were assisted by members the United States Forestry Service and the Federal Bureau of Fisheries.[103] Although estimates had placed the number of banker ponies as high as 6,000 as recently as 1935, there were fewer than 50 feral horses roaming the entire Outer Banks just three years later. On June 14, 1938, newspapers reported that eastern North Carolina's most expert marksmen had set out that morning armed with high-powered rifles and dum-dum bullets, in pursuit of the "final extinction of the banker ponies."[104]

To be sure, the feral horses on the Outer Banks had their defenders. In 1938, Mary Farrow Credle, who lived in the nation's capital but who held a tender spot in her heart for the Banker ponies, wrote an impassioned plea for people to stop killing the horses. "The ponies have never hurt the land of anybody and it is a crying shame that they have to be butchered to suit some fool theorist who has no practical knowledge what they have meant and can mean again to the folks in Eastern Carolina," she lamented.[105] As the feral horses of the Outer Banks dwindled in number, their origin myth likewise assumed new importance. Nearly every article written about the animals ascribed their origin to the earliest British settlers and the lost colony of Roanoke, though some suggested that these animals owed their origin to a shipwrecked Spanish galleon.

The horses on Cumberland Island experienced a different sort of fate altogether. Not satisfied with the number of feral horses who roamed their island, the Carnegies imported additional specimens in 1921. Aspiring archaeologist Oliver Ricketson (the twenty-five-year-old son of Margaret Carnegie Ricketson) had recently traveled to Arizona on his first professional expedition. While he was out west, Ricketson observed vast droves of feral horses. He procured a boxcar-load of these animals, "three- and four-year-old mares unbroken," and sent them back home to be released on Cumberland Island.[106]

While the range remained almost exclusively open throughout most of the Florida peninsula, feral horses were fewer in number than ever before. In 1934, the New Deal administrators tried to stamp out the last pockets.[107] The Federal Relief Administration employed "jobless horsemen" to round up the state's feral horses and "break" them for domestic service. Once broken, federal officials promised, the horses would be given to the state's most destitute people.[108] According to A. J. Ward, regional supervisor for the Rural Rehabilitation Division in the South, feral horses were still "plentiful in the northwestern and north central sections of the state." This is in the general vicinity of Bartram's "Great Alachua Savannah," also known as Paynes Prairie. What is more, he added that "droves of them have been hiding in the brush and feeding on the northern edges of the Everglades since pioneer days."[109] Some locals were skeptical, however. Ruth Parker, a self-described "old-timer" who had been born and raised in Fort Myers, Florida, doubted there were any feral horses left to be rounded up. "I have been all over the Everglades for weeks at a time and have never seen any wild horses," she wrote to Governor David Sholtz. "This is another money scheme. Somebody will make some money out of it but will not get a wild horse out of it. The Indians say there is no wild horses in the Everglades. I ask Cypers Billie yesterday if he ever sees any wild horses and he said no there is no wild horses in Everglades."[110] When biologist R. DeWitt Ivey conducted a survey of the mammals around Palm Valley in northeast Florida in 1959, he reported that feral horses had "only recently been exterminated."[111]

Finally, we turn to dogs, who were subject to unique pressures quite distinct from those shaping pigs and horses. The urge to celebrate domestic dogs intensified dramatically during the second half of the nineteenth century, when the "infrastructure of breeding was applied to dogs."[112] Wealthy people in Britain developed an abiding interest in "purebred" dogs, whose morphological variety they first publicly exhibited in 1859. Within five years, they were holding competitive bench shows that attracted thousands of dogs and as many people.[113] It was not long before wealthy people in the United States developed a similar fancy for fancy dogs. The phenomenon was most pronounced in the nation's largest cities, most of which were located in the Northeast and the Midwest.[114]

One might think that conditions were drastically different in the South, which boasted far fewer people and far smaller cities. In 1900, for example, none of the five southeastern states ranked among the top ten most populous in the nation. The region's largest cities at the time, Atlanta, Richmond, and Charleston, were the 43rd, 46th, and 68th most populous cities in the nation, respectively.[115] Even so, southern cities grew in dramatic fashion during the late nineteenth and early twentieth centuries.[116] The growth may pale when compared to other cities in different regions at the time, but when comparing this era of southern history to previous eras, the changes were profound.[117] Between 1870 and 1930, the region's human population increased from 4,300,000 to more than 11,700,000. During that same span, the region's various largest cities expanded dramatically. The growth was not the result of people migrating into the region from elsewhere, but rather because an increasing number of people already living in the South chose to abandon the rural life and migrated toward nearby cities and towns.[118] In 1870, less than 10 percent of the population qualified as urban. By 1930, more than 30 percent qualified. In Florida, more than half of the population was urban.

No other city in the South grew as much or as big as Atlanta in the decades following the end of the Civil War.[119] While barely 20,000 people lived in Atlanta in 1865, the city's population had tripled by 1890. Robert L. Adamson, editor of the *Atlanta Constitution*, confirmed the transition from "country hamlet into a metropolitan city."[120] Concordant with the rising number of people, there was also rising interest in dogs. Adamson was among the first to celebrate Atlanta's finest "purebred" dogs. He reported a wide variety of fine canids in the city, including greyhounds, pointers, setters, bloodhounds, pugs, mastiffs, and several St. Bernard dogs.[121] In 1900, the city's wealthiest citizens organized the Atlanta Kennel Club to serve local dog fanciers. Now that Atlanta was developing into a major metropolis, many of its citizens wanted an opportunity to showcase their dogs. That spring, the Atlanta Kennel Club hosted its first bench show (Figure 15). More than 300 dogs, many of them from the Northeast and the Midwest, entered into competition, and all manner of morphological variation was on display.[122] The *Constitution* reported that many of these dogs were valued at more than $5,000.[123]

Despite these prize-winning specimens, however, the vast majority of Atlanta's dogs had always been free-roaming canids whose reproduction was not closely monitored and whose wanderings were not generally restricted. They did not fare well during the Civil War, and some accounts suggest that Sherman's troops killed more than 500 dogs per day while marching through Atlanta. Despite this bloodshed, Georgia native Addison M. Weir recalled that thousands of dogs remained in Atlanta after Sherman's departure, and the city soon became known as a "wild dog region." Weir recalled that these feral dogs hunted in packs

Figure 15 Atlanta Kennel Club's first dog show, 1900
The Atlanta Kennel Club hosted its first bench show in 1900. The dogs on display boasted extreme morphological diversity, ranging from Great Danes to Chihuahuas.
**Source** – "Bench Show of the Atlanta Kennel Club..." *Atlanta Constitution*, May 9, 1900, 5.

like wolves, and that they targeted horses as often as sheep. He had not bothered to take a census, but estimated that "at least 45,000 dogs" roamed the city following the end of the war. Conditions had apparently not improved by the following decade, when municipal leaders employed a dogcatcher to round up any free-ranging canid not wearing a city-issued tag. Impounded animals were held in a shed on the outskirts of town for 24 hours. If no human came to claim them, the animals were summarily executed. The city's dogcatcher killed more than 350 in June 1879, though he assured concerned citizens, without a hint of irony, that 90 percent of these dogs he killed "belonged to the poorer classes of the city."[124]

The dogcatcher was only employed on an as-needed basis, and had evidently been out of service for some time by 1892, when the police captain complained that dogs were "taking over the city." Free-roaming dogs were even more conspicuous by 1896, when city administrators complained that dogs had been "allowed to roam at will" for far too long. The following year, Mayor Charles Collier issued an ordinance directing that every dog in the city had to be registered. "Dogs not registered will be lynched without further ado," officials warned.[125] Most citizens were resolutely opposed to paying a tax on dogs, and many (probably most) brazenly ignored the law. This put the dogcatcher in a difficult, and sometimes dangerous, position. On one occasion, an incensed citizen destroyed the dog wagon with an axe, thereby allowing twenty dogs to escape. Two years after that, another citizen broke into the pound and released fifty imprisoned dogs. On yet another occasion, still more citizens overturned the dog wagon, thereby allowing more than thirty dogs to escape.

Some of these episodes highlighted the South's strained racial tensions. In 1900, there were more than 3,300,000 African Americans living in the five southeastern states (compared with more than 4,400,000 European Americans and fewer than 6,000 Native Americans).[126] Like everyone else at the time, these African Americans increasingly abandoned the rural life in favor of the urban one. Many of them moved to cities outside the region, but many of them moved to cities within the South.[127] The number of African Americans living in Atlanta increased from roughly 9,000 in 1880 to more than 35,000 in 1900.[128] During the summer of 1899, many of them joined in protest against the city's dogcatcher. According to reports, it all began when two young "white" boys who worked on the dog wagon lassoed a "mongrel cur" on Auburn Avenue. It turned out the dog was owned by Carrie Thomas, a "respectable negro woman," who claimed that the two young boys had snatched the dog from her yard. Several witnesses corroborated her story, and by the time police arrived, more than a hundred of her neighbors had converged on the scene. "If that was a rich white man's dog," one protester complained, "the dogcatcher daren't touch it."[129]

While some accused the dogcatchers of being overzealous, others complained that they were not zealous enough. Citizens generally cited one of two different reasons when advocating for the eradication of free-roaming dogs. First, many cited the threat of rabies, also known as hydrophobia, a zoonotic disease all too common among stray dogs and easily transferable to humans.[130] The reader may recall that one of the greatest heroes in American literature, Atticus Finch, was compelled to shoot a "mad dog" in Harper Lee's legendary novel *To Kill a Mockingbird*. It may have been a sin to kill a mockingbird, but killing a rabid dog was just good sense. Second but no less often, citizens complained that free-roaming dogs barked, bayed, and howled all night long. Some citizens took matters into their own hands, scattering poisoned meat around their neighborhood. After one such attack, citizens awoke to find dead dogs littering the streets.[131] This tactic was not without risk, though. On one occasion, C. C. James brought suit against J. C. Evans for planting poisoned beef that was later eaten by James's "very fine fox terrier." This lattermost episode highlights an important point. Many people defended dogs, but only certain *kinds* of dogs. In response, the city's dogcatcher began separating "high-class dogs" from mere mutts. Dr. H. G. Carnes, representing the local humane society, endorsed the city's policy toward dogs, which "takes away numbers of stray dogs that are good for nothing on earth."[132]

In 1908, the Atlanta Humane Society (which had been created thirty-five years earlier) proposed a new method for disposing of excess dogs. Its members insisted that it would be more humane to asphyxiate the animals using poison gas and then cremate the bodies, but efforts stalled as members of the Humane Society and the city government tussled over fiduciary responsibility. Dogcatchers were still shooting strays wholesale in 1921, when the Atlanta Humane Society assumed control of rounding up the city's stray dogs. That June, the society opened a state-of-the-art dog pound on Marietta Street, modeled after the dog pound in Boston. The society promised that strays would be kept in clean pens, and that unclaimed animals would be put to death by "the painless method of electrocution."[133]

Because there was no dog census taken, it is impossible to know the exact size of Atlanta's dog population, to say nothing of its composition.

Nevertheless, bits of evidence suggest that the number of dogs living in the city was quite large. During the summer of 1903, the city clerk reported that 888 dogs had been tagged and registered that year, and that roughly 500 dogs had been executed. Both of these numbers represented small fractions of the actual population. When city officials were asked to estimate the number of dogs living in Atlanta at the turn of the century, their guesses varied all the way from 10,000 to 40,000. The city's dogcatchers impounded more than 2,000 dogs in 1920 and more than 4,000 dogs in 1935. Three years later, the latest dogcatcher, William H. Browne, reported that approximately 6,000 dogs entered the pound each year, and that only 10 percent were reclaimed. Moreover, he confirmed that virtually none of the dogs he killed were purebred. "A lot of people who came here were looking for fine dogs. Some asked for dogs with pedigrees," Browne reported. "We don't have dogs like that. We just have plain dogs – dogs that play with the neighborhood gang, or just waifs. They are good dogs, but they don't have any background."[134]

Similar patterns were repeated in other southeastern cities. Richmond was still the largest city in the South when it hosted its first dog show in 1879. Recognizing that most dog fanciers lived in the Northeast, *Forest and Stream* advertised the show by providing railroad rates from New York to Richmond. As more and more northerners descended on the former capital of the Confederacy to compare canids, old rivalries began to melt away. Dog shows in Richmond routinely attracted hundreds of canine entries, and remained a mainstay in the city throughout the late-nineteenth and early-twentieth centuries.[135] Other bench shows in the region were less successful. When the citizens of Charleston attempted their first dog show in 1891, the event garnered just sixty-five entries. Participants openly complained that the paltry competition was hardly worth the long trip from the Northeast. "The Charleston people, while anxious to promote a 'dogly' interest in the South, hardly go the right way about it," one northern visitor complained.[136] Complicating matters, people who visited Charleston with "blooded dogs" might not leave with the animals. In 1901, the *Sunny South* confirmed that "blooded dogs are in high favor in Charleston," and that this demand had led to a rash of crime. "Many thoroughbred dogs have been stolen from ships in port at Charleston in recent years," the newspaper reported.[137]

Charleston's troubles did not dissuade the region's other burgeoning cities from hosting similar events. During the first two decades of the twentieth century, Charlotte, Columbus, Augusta, Columbia, Camden, Pinehurst, and Norfolk (to name just a few) all hosted dog shows.[138] Yet all of these cities also contained feral dogs. Reporters dubbed Richmond a kind of "dog heaven," and calculated that the city contained more than 10,000 free-roaming dogs "without home or mother." When members of the City Council debated how they should address the city's large and growing plague of "worthless curs" that July, editors at the *Dispatch* suggested wholesale eradication. When one reporter visited the pound in 1902, he found that the dogcatcher kept purebred animals longer, on the grounds that they were more valuable and thus more likely to be purchased. Without question, however, the most numerous dogs at the pound were genetically undifferentiated mutts who had never known human companionship. The dogcatcher had collected hundreds of these animals over the previous two months, the vast majority of whom were gassed and cremated without further ado.[139]

Charleston also boasted a large number of free-roaming dogs at the turn of the century. According to some (probably exaggerated) estimates, the number of dogs in the city exceeded the number of humans. "It is astonishing to behold the number of homeless curs in Charleston," one citizen remarked during the summer of 1908. To address the problem, city officials prohibited any dog, tagged or otherwise, from appearing in public without a human escort. Because so many passionate pet owners failed to respect the dogcatcher's authority, the city's police chief assigned one of his officers to accompany the dog wagon. Although all dogs were technically subject to the new law, reports explained that the dogcatcher would direct his efforts toward exterminating homeless "mongrels."[140]

Other southeastern cities also struggled to deal with feral dogs. When W. L. Creath rattled off a letter to the editors of the *Tampa Tribune* in 1911, he decried the large number of "filthy" dogs who roamed Tampa, and he was unequivocal in his condemnation. He urged the city hire a dozen dogcatchers to kill thousands of the animals post haste. Meanwhile, conditions were no different in nearby Ybor City, which he likened to "hundreds of small dog farms." Creath was adamant that dogs should be domesticated in the fullest sense of the word, meaning they should never be left out of the house.[141]

Meanwhile, citizens of Greensboro had grown weary of the large numbers of mutts roaming their city by 1918. "There is a criminal class of dogs as there is a criminal class of human beings," one citizen explained in a spirited letter to the editor. He exempted the "self-respecting gentlemen and lady dogs" from his scorn, but campaigned vigorously to exterminate "tramp dogs." Tellingly, he drew explicit parallels between dog control and human eugenics, writing that the city should "protect good dogs, kill bad ones, and control the dog population just like men try to control the human being procession." Meanwhile, another Greensboro man devised a different method. J. L. Jordan owned and operated the Dan River Kennel Club, which advertised "highly pedigreed dogs" for sale at $60 each. He received orders from across the nation, but instead of sending the promised purebred specimens, he sent "worthless neighborhood curs." When the ruse was discovered, Jordan was charged with mail fraud in federal court, and was eventually fined $600.[142]

Accurately estimating the number of dogs in the South during this era is well-nigh impossible, especially since so few people registered their dogs. Consider the following statistics. In 1906, a farmer in Clifton Forge, Virginia, confidently asserted that people pay taxes on well below half of the dogs in the commonwealth. He probably overestimated. Despite complaints about the dog's ubiquity in South Carolina's biggest city, for example, the citizens of Charleston (numbering approximately 55,000) paid taxes on just 454 dogs in 1905, the smallest number in the state. In 1914, D. W. Thomas, tax collector for Spartanburg, estimated that the town probably contained more than 3,000 dogs, but that only 61 of them had been tagged and registered. Compliance was only slightly higher in Macon, where officials were confident that 30 percent of the city's canine population had been tagged and registered in 1915. One year later, tax collectors in Wilmington estimated that their town contained more than 2,000 dogs, but that only 700 had been registered. In 1919, citizens in Columbia registered more than 600 dogs, but this number did not include the "hundreds of worthless dogs (who) roam the streets at all hours of the day and night."[143]

In addition to the large number of dogs who either cohabited with people or who lived on the streets of southeastern cities, dogs sometimes turned up in the woods as well. In 1887, hunters discovered a "den of wild dogs" in the rugged mountains near Asheville, North Carolina, and newspaper accounts described the animals as "ferocious."[144] Just one year later, people in mountainous Wilkes County chanced upon a litter of puppies found living in a hollow log. Observers described the animals as "exceedingly fierce" and "very wild."[145] In 1895, people discovered "wild dogs" in the Pine Barrens near Vienna, Georgia. Newspapers reported that the animals resisted socialization, and that they were just "as wild as ever."[146] In 1899, there were reports that packs of "wild dogs" terrorized the margins of human settlement near Wilmington, North Carolina.[147] Meanwhile, in the summer of 1907,

Henrico County, Virginia, was "overrun with wild dogs" who wreaked havoc on the county's livestock. Tellingly, one candidate for public office readily admitted to killing more than 75 dogs.[148] Meanwhile, in one vast stretch of woods southwest of Raleigh, packs of "wild dogs" persisted for more than thirty years. Employees at the nearby Hospital for the Insane waged constant war against these dogs, but the animals proved almost impossible to eradicate.[149]

*Chapter Summary*

In the decades following the Civil War, southerners became increasingly entrenched in national and international markets. Although they had already imported enormous numbers of pigs into the region prior to the war, they imported even more after the war. As human population densities increased, and the number of southerners relying on local swine for sustenance decreased, support for the open range began to vanish. Between 1862 and 1949, the southern range slowly but surely collapsed in piecemeal fashion. Theoretically, closing the range should have spelled certain doom for the region's feral hogs. After all, swine were technically not allowed to exist anywhere save for inside an enclosure. Huge numbers of these tenacious survivors nevertheless managed to carve out an existence in the southern wilds. These animals did not go unnoticed, however. As tourist sportsmen from the Northeast began descending on the South in droves, they often marveled that the region still contained feral hogs. It did not take them long to designate these hogs as "game" to be pursued and destroyed. This was the origin of recreational pig hunting in the South, an enterprise that has since had profound evolutionary consequences for the region's porcine genome.

Meanwhile, the number and value of horses increased dramatically during the latter half of the nineteenth century. The number in the South also increased, but not as dramatically as the rest of the nation. That is because southerners were growing ever more reliant on imported mules. By the turn of the century, mules accounted for more than 40 percent of the region's equids. All the while, a few relict bands of feral horses remained scattered along the sparsely populated barrier islands that dotted the southeastern seaboard. And then something happened: Death arrived in the form of an internal-combustion engine. When the number of automobiles skyrocketed during the first few decades of the twentieth century, the implications for the horse were lost on no one. Beginning in the 1920s, the value and number of horses began to plummet. By the late 1930s, the nation's equine population was a fraction of its former size. As for feral horses, the invention of the automobile meant different things for different populations. Some were romanticized and commercialized more than ever, and some were gunned down by the hundreds because of their purportedly negative ecological impact.

Finally, the distinctions between domestic and feral dogs grew more pronounced, and thus even more evolutionarily significant, during this period. On the one hand, some dogs were prized for their tightly controlled genetic pedigree. They were often purchased for thousands of dollars, and paraded before discerning judges at competitive "bench shows." These dogs were not only encouraged to reproduce, but most were themselves the product of carefully arranged couplings. This phenomenon was most acute in the Northeast and the Midwest, but it was also present in many of the South's largest urban centers. Despite these prize-winning specimens, however, the vast majority of dogs in the South remained genetically undifferentiated. Most of these strays, or curs, lived on the periphery of human habitats, and were not specifically bonded to any particular human companion. These dogs were actively discouraged from reproducing, and untold millions were slaughtered on an annual basis.

# 6

## EVERYTHING IN ITS RIGHT PLACE
### WILD, DOMESTIC, AND FERAL POPULATIONS IN THE MODERN SOUTH

*The past is never where you think you left it.*
~ Katherine Anne Porter

Prior to 1970, the southeastern states contained fewer immigrants than any other region in the United States.[1] Thanks to the expansion of the interstate highway system and, more importantly, the invention of air-conditioning, millions of people have immigrated into the South over the past half-century.[2] According to the 2010 federal census, the five southeastern states are now home to more than 50,000,000 people. Among this number, approximately 34,000,000 people self-identify as "white," while 10,000,000 self-identify as "black or African American." Meanwhile, 273,000 people self-identify as "American Indian." This means that the Native American population has grown dramatically since the population's nadir in 1900, but obviously still represents a comparatively small portion of the region's human population.[3] For comparison's sake, more than 1,400,000 people self-identify as "Asian." Another 1,400,000 self-identify as "more than one race," a number that is, in this author's opinion, *well* below the actual number given humanity's well-documented proclivity toward admixture.[4]

When we last discussed the South's domestic equids, their numbers were plummeting. Across the region (and the world), millions of people were making the switch from animate to inanimate modes of transportation. Though horses and mules had served alongside people in the region for centuries (and alongside humans on Earth for millennia), they were, within the span of a generation, utterly superfluous. They no longer provided any real advantages when it came to transportation, hauling, agriculture, or war. Consequently, their population shrank in dramatic fashion. More than 1,300,000 mules had lived in the five southeastern states in 1920, but trans-Appalachian mule-production business had otherwise collapsed. There were still 520,000 mules in the South in

1959, when census takers counted mules as a distinct category for the last time. These creatures owed their presence in the region to longevity more than utility, but, in due course, they too fell by the wayside. Just like that, an animal who was once synonymous with the South disappeared entirely, harkening Herodotus' ancient observation that mules are cursed with "no ancestry and no hope of posterity."[5]

Those horses who were not siphoned into mules likewise experienced a major population contraction. Whereas the agricultural census had counted more than 700,000 horses in the five southeastern states in 1920, they counted fewer than 120,000 in 1974.[6] Unlike mules, however, horses retained their fertility and could thus reverse the population nosedive. After bottoming out around 1974, the region's horse population crept steadily upward for the next few decades. When the USDA conducted the agricultural census in 2007, federal authorities counted more than 400,000 horses in the region. To explain this increase, researchers cited several factors, including the expansion of suburban communities into former pastureland, an expanding middle class that could afford to keep horses as pets or companions, and the South's age-old appetite for gambling on horse racing.[7]

The region's largest concentration of horses resides in northern Florida.[8] This is not some remote wilderness where the locals have rejected automobiles. On the contrary, this is an area where people direct highly intensive breeding operations aimed at producing the fastest racehorse in the world. These efforts began in earnest in 1943, when Midwestern businessman Carl G. Rose established the region's first thoroughbred breeding farm in Marion County, Florida. He predicted that the region's calcareous soil and favorable climate would help produce strong, fleet thoroughbreds. Whatever the reason, his stock began producing reliable winners, causing other horse fanciers to descend on north-central Florida.[9] The region's equine operations gained international renown in 1956, when a horse by the name of "Needles" became the first Florida-born thoroughbred to win the Kentucky Derby and the Belmont Stakes. He also placed a close second in the Preakness, thus narrowly missing the Triple Crown.[10] Following Needles's success, breeding operations in the region expanded dramatically. Marion County has since produced another five Kentucky Derby winners, and is now home to hundreds of thoroughbred farms.[11] (The fastest racehorse in recorded history, Secretariat, was born and raised on Meadow Farm in Caroline County, Virginia.[12]) Always popular in the South, horseracing is now a multibillion-dollar industry and an increasingly fine science. As of early 2016, scientists have not yet used the newly discovered CRISPR-Cas9 gene-editing technology to manufacture faster racehorses, but they *have* identified a "gene for speed" in thoroughbred horses, and so it is only a matter of time.[13]

Those horses who remain domesticated are now subject to more intense anthropogenic selection than ever before, but that has earned them no guarantees. Many geneticists are convinced that thoroughbred horses have been "overbred," and that they have far more potentially harmful mutations than ancient horses.[14] Meanwhile, the number of domestic horses in the South has trended upwards over the past forty years or so, but its growth is not inexorable. In fact, the USDA's most recent agricultural census shows that the number of horses in the South once again *decreased* between 2007 and 2012.[15] Experts cite two major reasons for this population contraction. First, the financial meltdown that began in late summer of 2008 has left people in the South with less expendable income. "State, local government, and animal welfare organizations report a rise in investigations for horse neglect and more abandoned horses," the US Government Accountability Office recently reported.[16] Second, the federal government has enacted tighter restrictions on horse slaughter, which has dramatically reduced the demand for horses. This sounds counter-intuitive, but makes sense when one realizes that processing facilities in the United States slaughtered more than 2,400,000 horses during the previous three decades.[17] Without financial incentives, traders breed fewer horses every year. In 2001, about 13 percent of the nation's horses were less than one year old. By 2011, just 3 percent were less than one year old.[18]

With so little demand for domestic horses, one might think that the demand for feral horses would have dried up altogether. And yet that has not proven true. There are still about 500 feral horses subsisting in scattered pockets across the South. Among these, the Chincoteague horses (who now exist solely on Assateague Island) are the most famous. In 1943, the federal government established the ChincoteagueNational Wildlife Refuge on the southern end of AssateagueIsland. Government officials allowed the feral horses to remain, and they allowed the Fire Department to continue the annual roundup, with the stipulation that the horses should not exceed approximately 150 in number. The annual roundup continued to attract thousands of visitors every year, though attendance received an extra boost in 1947, when Marguerite Henry published her award-winning children's book, *Misty of Chincoteague*.[19] The book repeated the fable that horses first arrived on Assateague when a Spanish galleon wrecked against the reef centuries earlier. The feral horses on Assateague grew even more famous in 1960, when Henry's book was adapted into a major motion picture. The following summer, a record crowd of more than 50,000 people attended the annual pony-penning festival in Chincoteague.[20]

Although it has not factored into our narrative thus far, the reader should know that the southern part of Assateague Island lies in Virginia, and that the northern part lies in Maryland. This invisible line had never previously prevented horses from roaming the entirety of the island, but that all changed in the 1960s. In quick succession, the federal government bought the Maryland side of the island from beleaguered investors, established Assateague Island National Seashore, and erected a wire fence that would prevent Virginia's horses from coming over. At the same time, the National Parks Service removed all of the privately owned horses from the Maryland area. Given a clean slate, as it were, local Jaycees donated 21 horses to form the nucleus of a new herd.[21] By 1968, there were 28 horses living on the Maryland side of the island.[22] Park officials adopted a hands-off approach to managing these animals. They did not round them up and they did not auction them off. By 1979, there were more than 60 horses on the Maryland side. They were still "essentially in quarantine" in 1985, by which point biologist Ron Keiper urged action. "Control of the population is, or soon will be, needed," he wrote.[23] Toward that end, biologist Jay Kirkpatrick devised an ingenious method of immuno-contraception in the early 1990s. In simplest terms, he shot the horses with birth-control darts. More recently, biologists analyzed genes from the Maryland herd and found that the animals are at risk of becoming inbred. To ameliorate this situation, they write, the Park Service will be compelled to feralize additional horses at some point in the future.[24]

By comparison, the Chincoteague Fire Department has consistently tinkered with the feral horses on the Virginia side of Assateague Island, and this has necessarily influenced the herd's evolution. Foals are annually rounded up and auctioned off, so the animals who remain behind are older than would otherwise prove true. It has also accelerated reproduction. Because fewer offspring are being weaned at any given time, mares are more likely to get pregnant and have more foals. Even more significantly, the horses on the Virginia side have also been subject to periodic genetic interchanges. This was true long before the Fire Department assumed control, and it has remained true after.[25] When a category-3 hurricane killed half of the island's horses on Ash Wednesday, 1962, locals responded by feralizing still more horses. Henry celebrated this restocking event in *Misty*'s sequel, *Stormy*.[26] In 1978, the firemen were compelled to euthanize many of the horses who had recently tested positive for equine infectious anemia. To supplement the suddenly smaller herd, they procured forty feral mustangs from California. Most of these transplanted mustangs failed to adapt to harsh island living, however, and died within the first year.[27]

While the Fire Department has not introduced any new horses since 1978, they have carefully managed the existing herd's health. Doing so has helped sustain the community's economic viability.[28] The Chincoteague pony-swim remains as popular as ever, annually attracting thousands of visitors to Virginia's most remote islands. When the Fire Department held its annual roundup in July 2013, they auctioned off 57 foals. The young horses sold for an average of $2,000, though one exceptionally attractive three-month-old female was purchased for more than $12,000. The high-bidder purchased the horse as a birthday present for his octogenarian mother, who was also there in attendance. The new owners turned that young horse loose on Assateague, so that she might remain part of the herd's genomic future.[29] Despite these well-documented introductions, the shipwreck myth persists. For example, the Visitor Center at the Chincoteague Wildlife Refuge features a dramatic mural that depicts horses scampering ashore during a storm while their ship founders behind them (Figure 16).[30]

Humans have also influenced the various bands of feral horses that live in the Outer Banks of North Carolina. The smallest of these populations lives on Ocracoke. Though geographically isolated, these horses have nevertheless been subject to several well-documented genetic influxes over the years.[31] These introductions notwithstanding, the horses were still free-roaming in 1954, when Major Marvin Howard founded the most unusual Boy Scout troop in the nation. Each boy in the troop was responsible for catching and riding a feral horse of his choosing.[32] These mounted Boy Scouts garnered national attention in 1956, when they were featured in *Boy's Life*. This brush with domesticity was a sign of things to come. When the federal government announced plans to establish Cape Hatteras National Seashore on the island, state officials took steps to tidy up the joint. In 1957, transportation officials opened a paved road (Highway 12) that connected Ocracoke to the ferry terminal on Hatteras Inlet. To make the road safer for tourists, the state legislature passed a law that explicitly prohibited livestock from running free or at large on Ocracoke. Crucially, they exempted those horses under the

118   *Feral Animals in the American South*

Figure 16  Mural depicting fabled origin of Assateague Horses
Located in the Chincoteague Wildlife Refuge Visitor Center on Assateague Island, this mural depicts the purported origin of the Assateague horses. Despite numerous well-documented genetic introductions over the years, the shipwreck myth survives.
Photograph of mural taken by the author.

care of the Boy Scouts, provided the horse population never exceeded thirty-five in number.[33] To protect them from high-speed automobiles, the scouts built a fence to enclose the animals. Completed in 1959, the enclosure was three miles long and a half-mile wide.[34]

Technically, these animals ceased to be feral, inasmuch as they ceased to be free-roaming, but that did not mean they were saved. By the late 1960s, observers noticed that the Boy Scouts had outgrown ponies and "the present crop of Scouts has other things on its mind."[35] The herd had dwindled from nearly seventy in 1956, to just twenty-five a decade later. Karl T. Gilbert, superintendent of Cape Hatteras National Seashore, saw no reason to keep them. "Records now show that the 'Ocracoke ponies' are nothing more than run-of-the-mill horses," he wrote, adding that well-documented genetic influxes had eliminated the horses' historical significance. "The true 'Ocracoke pony' has long ceased to exist," he wrote.[36] By 1976, there were just nine horses left on Ocracoke. Around that time, the National Parks Service made a special effort to save the herd from extinction. Officials imported an Andalusian stallion named Cubanito to procreate with the remaining mares on hand. The herd gradually increased, and there are now about twenty-five horses on Ocracoke. Provided with food and medical attention, these animals are a living diorama of *re*-domesticated creatures.[37]

A second population of horses, currently numbering between 110 and 130, lives on Shackleford Banks.[38] Although Fred Olds had reported 3,000 horses in the area in 1925, there were fewer than 125 horses roaming the island by 1952.[39] "North Carolina's famed Banker Pony is slowly disappearing," Norwood Young reported at the time.[40] To save the herd from extinction, the aforementioned 1957 law that allowed the Ocracoke horses to remain also made an exception for the horses on Shackleford Banks.[41] Nevertheless, their numbers continued to drop. In 1960, Thomas W. Morse, Superintendent of State Parks, reported that there were only 20 horses left on Shackleford Banks. According to Morse, the horses had to compete with hundreds of cattle, sheep, goats, and pigs for sustenance.[42]

When the federal government established Cape Lookout National Seashore (which encompasses Shackleford Banks) in 1966, officials removed all of the livestock from the island save for horses. Since then, the horses have been largely isolated and the population has continued to rise.[43] By 1982, there were more than a hundred horses on Shackleford Banks. By 1996, there were approximately 240 horses on the island.[44] In response to this perceived overcrowding, the Park Service rounded up the horses to give them physical health inspections. Seventy-six of the animals tested positive for equine infectious anemia, and were subsequently euthanized. There was a loud public outcry, which eventually resulted in federal legislation. In 1998, President Bill Clinton signed a law guaranteeing the feral horses' right to exist on Shackleford Banks, provided they number no fewer than 100 and no more than 110.[45]

The third and final horse population in the Outer Banks is located in the northern end of the islands, near Corolla, and currently numbers between 90 and 120.[46] This area was developed much later than the rest of the Outer Banks (Corolla did not even get electricity until 1955), and so there is less documentation regarding the historical distribution of feral horses in the region.[47] We know there were dozens roaming the northern banks when members of the Spanish Mustang Registry visited Corolla in 1982. When six horses were killed following a single automobile accident in 1989, concerned citizens of Currituck County (where the horses were located) established the Corolla Wild Horse Fund.⁴⁸ Since then, these individuals have tirelessly sought to protect Corolla's feral horses. The organization assures visitors to its website that the horses are descended from Spanish mustangs, and touts endorsements from the Spanish Mustang Registry and the American Livestock Breeds Conservancy attesting as much. According to the *Official Horse Breeds Standards Guide*, the feral horses of Corolla are among the "purest herds" of Spanish mustangs in existence.⁴⁹ This is more than just idle boasting about fabled pedigrees. If the animals are descended from Spanish stock, they may well represent a unique genetic reservoir and gain federal protection accordingly. This could be vitally important to the herd's continued existence. Unlike the herds in Ocracoke and Shackleford Banks, the horses on Corolla, who continue to roam both public and private lands, have received no special assurances from the federal government. Somewhat disconcertingly, recent genetic analyses suggest the Corolla herd is *not* descended from Spanish stock. The horses' genetic amalgamation and purportedly negative environmental impacts cast considerable uncertainty on the herd's future.⁵⁰

Farther south, more than a hundred horses still roam Cumberland Island off the coast of southern Georgia. This herd had already been subject to numerous well-documented genetic introductions by 1972, when the federal government established Cumberland Island National Seashore. The National Park Service (NPS) allowed the horses to remain on the island, and they even allowed a nearby resident to introduce several Arabian horses to the herd in 1992. There are currently about 140 horses on the island. For the past decade or so, the foaling rate and the death rate have remained relatively close to one another, meaning park officials have not had to remove any horses during that span. The NPS continues to allow feral horses to roam Cumberland Island, though it should be noted Park officials also provide a laundry list of items (dunes, wetlands, water quality, ground stability, historic structures, archaeological sites, native grasses, and wildlife) that the free-roaming horses purportedly threaten.⁵¹

Numerous herds have also been deliberately reintroduced to places where feral horses were known to roam in the past. For example, the reader will recall that William Bartram observed huge droves of feral horses in the Great Alachua Savannah in the late eighteenth century, but that the federal government swept the last of these animals off the range in the mid-1930s. Well, many citizens felt that something meaningful had been lost when the last of Florida's free-ranging horses disappeared. In 1985, local horse enthusiasts donated six "cracker horses" (the colloquial name for the formerly feral landrace unique to Florida) to Paynes Prairie State Park near Gainesville. There are now about 30 horses in the park, which officials have decided is more than enough.[52] Meanwhile, there are approximately twenty horses known as "marsh tackys" (the colloquial name for the formerly feral landrace unique to the Lowcountry) now roaming Little Horse Island off the coast of South Carolina. Citing the animal's lack of cultivation over the years, breed-standard guides refer to the marsh tacky as a "time capsule of genetic material from colonial times."[53] Sure enough, legend holds that these animals are descended from the free-ranging swamp horses whom Francis Marion conscripted into service during the Revolutionary War. That may well be true, but they are more recently descended from horses who were deliberately released onto the island in the late 1950s.[54] Finally, about fifty horses from Assateague were introduced to Grayson Highlands State Park in southwestern Virginia in 1975, and their descendants continue to roam the mountainous terrain all these years later.[55]

Finally, in addition to domestic and feral populations, the South now also contains numerous "wild" horses as well. These animals were first discovered in 1880, when Russian Colonel Noklai Michailovich Przewalski led an expedition into the Altai Mountains of Mongolia. These free-ranging equids (thereafter known as Przewalski's horses) were celebrated as the last truly wild horses, or, more precisely, the domestic horse's closest wild, never-domesticated cousins. Over the next few decades, animal dealers from across the world ventured into the region to capture these animals. The equids were whisked away to zoos in New York, Philadelphia, and Washington. By 1969, they were extinct in the wild.[56]

Przewalski horses first arrived south of the Potomac River in the 1970s, when the National Zoological Park moved its wild equids to the newly acquired Front Royal campus in northern Virginia. Scientists at the Smithsonian Conservation Biology Institute have since worked diligently to preserve these equids. Though there were once fewer than 20 Przewalski horses left on the entire planet, there are now more than 1,200 individuals located at zoos and breeding centers around the world.[57] In the summer of 2013, Smithsonian officials announced that a Przewalski horse had been born on the Front Royal campus, the first time a "wild" horse had been produced via artificial insemination.[58]

Before reflecting on the distribution of extant equids, let us first account for our other two species. When we last discussed the nation's domestic pigs, the vast majority of them were coalescing in the Midwest. As a result, southerners of every race and gender were slowly but surely becoming ever more entrenched in international markets and gene flows. It should be noted, however, that this change did not happen overnight. As recently as 1918, swine were distributed rather evenly across the South, and pigs were still being produced on 75 percent of the farms.[59] In due course, however, things began to change, and pigs are now more centralized than ever before. The largest population remains concentrated in the Midwest, but the second largest population is now concentrated in eastern North Carolina (Figure 17).

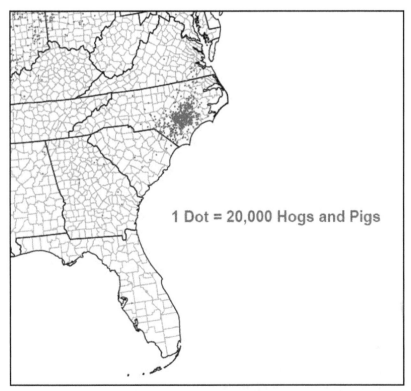

Figure 17 Distribution of pigs in the South, 2012
Almost all of the pigs in the South are concentrated in eastern North Carolina, where they live in total-confinement processing facilities.
**Source** – "Hogs and Pigs – Inventory," 2012 Census of Agriculture, National Agricultural Statistics Service.

Many farmers in North Carolina were still practicing free-range husbandry as recently as 1915. Things began to change that winter, however, when a catastrophic storm killed more than half of the pigs in the state. Thereafter, agriculturalists began to emphasize the importance of housing pigs instead of allowing them to go at large.[60] In 1918, the state appointed William Shay to head the state department of agriculture's new swine division.[61] Shay was an experienced Midwestern pig farmer who successfully promoted improved husbandry methods across the state. Despite his efforts, however, North Carolina was still not producing as many hogs as Georgia as recently as 1964, and the Tar Heel state's processors were still only operating at 70 percent capacity.[62]

That was about to change. In 1964, the Surgeon General's office conclusively established that smoking tobacco causes cancer. The report proved a boon to the nation's public health, but it cast considerable uncertainty on the future marketability of North Carolina's most profitable cash crop. Lawmakers on Tobacco Road understood that the state needed a new agricultural staple, and they settled on total-confinement pig farming.[63] There was already precedent for this type of agriculture in the South. Monica Gisolfi has recently described the origins of total-confinement poultry farming in mid-century Georgia, but the practice meant something different for pigs.[64] In "total confinement," pigs lived their whole lives in pens that scarcely gave them room to stand. That meant they would no longer waste precious calories on pointless behavior, like walking. The pigs were provided with food and water via an automated delivery system, and their waste was filtered through slats in the floor. North Carolina established the Swine Development Center near Rocky Mount in 1965, designed to help farmers in eastern North Carolina establish total-confinement facilities.[65]

The state's swine breeders received another boost in 1979, when lawmakers in Raleigh effectively rendered it impossible to sue any agricultural operation, and then again in the early 1990s, when zoning officials decided that massive hog-processing facilities fell within the legal definition of "farm." As a result, farmers who had previously produced hundreds of pigs per year now produced thousands per year. By the early 1990s, the state's pig industry surpassed one billion dollars annually.[66] By the early 2000s, total-confinement pig "farms" were the norm. Whereas the average pig farm contained just 9 pigs in 1940, the average pig farm contained almost 900 pigs in 2007.[67] But even those statistics do not capture the profundity of recent changes. In 2005, fully 82 percent of the nation's pigs were located on farms that processed more than 5,000 pigs per year. By 2008, that number had risen to 88 percent. Ten states now account for over 85 percent of the nation's total pig inventory.[68]

No other company better illustrates the extensive (and intensive) nature of modern pig husbandry than Smithfield Foods. Headquartered in southeast Virginia, Smithfield is the world's largest pork supplier and producer.[69] The company operates facilities in more than 20 states, but its largest plant is located in Bladen County, North Carolina. Established in 1993, this "farm," which slaughters more than 36,000 hogs *every single day*, is now the largest meat-processing facility on the planet.[70] Bladen County and its contiguous neighbors, Duplin and Sampson Counties, contain approximately 150,000 humans, but more than *5,000,000* pigs.[71] When footage that had been filmed inside the plant was featured in the documentary, *Food, Inc.*, Smithfield and other companies faced a wave of criticism. In response, the nation's leading meat producers are now trying to pass "ag-gag" legislation that would make it illegal to record what happens inside a slaughterhouse.[72] In other words, these corporations *really* do not want consumers to see the sausage being made. It is not enough that people never see pigs; they must also forget pigs exist.[73]

This kind of intensive selection has influenced porcine evolution in a number of profound ways. For example, meticulous breeders have increased the number of pigs in each litter, while decreasing the number of sows. This means they commit fewer animals to reproduction, yet still increase their annual crop of pigs.[74] Economic expediencies have also influenced the pigs' biogeographic distribution, as evinced by their incredibly dense concentration in North Carolina and their total absence from anyplace else. Furthermore, cultivating only the most desirable pigs has left the porcine genome more homogenous than ever, and has resulted in the "genetic erosion" (read: extinction) of hundreds of ancillary breeds.[75] And yet these evolutionary changes will likely pale in comparison to the changes that will be wrought by genetic engineering. Scientists have already used gene-editing tools like TALEN and CRISPR-Cas9 to demonstrate that they can specially design larger-than-normal pigs. For that matter, scientists have also used the same technology to produce cuddly "micro-pigs" whom they sell as pets.[76]

In the latest and least expected turn of events, a Chinese firm, Shuanghui International, recently purchased Smithfield for more than $4 billion. The move is especially surprising in light of the fact that China already contains far more pigs than any other nation on Earth.[77] A number of US Senators expressed concern that transferring ownership of the nation's pigs to Chinese authorities might render Americans somewhat vulnerable.[78] These fears notwithstanding, the Senate's Committee on Foreign Investment approved the transaction, thereby allowing the largest ever takeover of an American company by a Chinese counterpart.[79] What is more, Shuanghui probably got a steal. After all, the average American citizen eats more than 50 pounds of pork every year (the average southerner slightly more), and pork is the most widely eaten meat on Earth, despite the fact that many religions prohibit consumption. Indeed, scientists estimate that pigs account for half of humanity's daily protein intake.[80]

When we last discussed the South's feral pigs, their range was contracting. By 1920, the vast majority of the southeastern range had already been closed. A few of the mountainous counties in Georgia and North Carolina retained the open range for another twenty years, but they too eventually succumbed. The last great stretch of the southern range collapsed in 1949, when Florida passed a statewide law prohibiting livestock from roaming at large. Closing the range did not just make all the region's feral pigs disappear overnight, though. Many continued to carve out a niche in coastal lowlands and riverine shallows.[81] When biologists Jack Mayer and Lehr Brisbin counted the region's feral pigs in 1991, they reported that approximately 2,000,000 feral hogs had established populations in twenty states.[82] One might think that another few decades of control measures would have shrunk the nation's population of feral pigs by now, but nothing could be further from the truth. When they reassessed the nation's feral pigs in 2009, Mayer and Brisbin reported that their number and range had both increased dramatically. They estimate that 5,600,000 feral pigs have established populations in thirty-six states, and that the five southeastern states contain around 2,000,000 (Figure 18).[83]

Biologists and management officials are unanimous in their opinion that feral pigs are highly "invasive" animals who should be eradicated with extreme prejudice. This professional consensus dates further back than one might otherwise expect. As early as 1899, biologists were already decrying the ecological impact of feral animals. T. S. Palmer, Assistant Chief of the Bureau of Biological Survey, reported that domestic animals, like domestic plants, may "run wild and become so abundant as to be extremely injurious." Rather than referring to these wayward animals as "feral" or even "invasive," however, Palmer instead referred to them as "noxious."[84] In 1910, *Forest and Stream* had likewise disparaged the ecological effects of feral pigs, especially their rampant destruction of bird eggs.[85] When LeRoy Stegeman published the first scientific analysis of

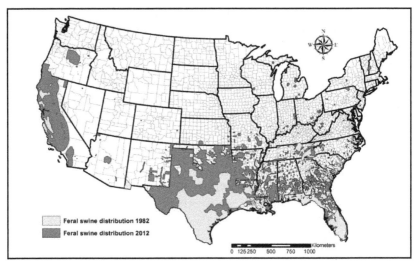

Figure 18 Map showing range expansion of feral pigs, wild boars, and hybrids
The number of free-ranging pigs living in the United States has exploded over the past thirty years, and, as a result, their range has expanded in dramatic fashion. Biologists blame hunters for not only translocating feral pigs, but also releasing wild (never-domesticated) boars in an attempt to shape porcine morphologies. Indeed, free-ranging pigs are now the second most popular hunting trophy in the American South, trailing only whitetail deer.
Source – Sarah N. Bevins, et al, "Consequences Associated with the Recent Range Expansion of Nonnative Feral Swine," *BioScience* 64 (April, 2014): 295. By permission of Oxford University Press on behalf of the American Institute of Biological Sciences 2014.

feral pigs in 1938, he included photographs that showcased the animals' destructive habits.[86] When ecologist Charles Elton published his landmark volume, *Ecology of Invasions by Plants and Animals*, he referred to feral pigs as "problem animals."[87] McKnight later confirmed the same.[88] Now that the feral pig population is growing rapidly, management officials are more insistent than ever that feral pigs are invasive pests.[89] They are routinely accused of destroying crops, spreading disease, damaging property, and even attacking people.[90] Biologists estimate that feral pigs cause more than a billion dollars in damages to crops in the U.S. on an annual basis.[91] The feral pigs in eastern North Carolina are proving especially worrisome, precisely because of their proximity to the industrialized pig "farms" in that region.[92] In 2014, Congress approved

$20 million for the USDA's National Feral Swine Damage Management Program in 2014, the USDA's first attempt to control feral pigs on a national level.[93]

Biologists and management officials use a variety of methods to kill feral swine, but most of them involve a gun. One can stalk the animal, but this method can only meet with a certain level of success. Feral pigs are elusive and they are mean, and both qualities complicate their eradication. Some biologists have found greater success using night-vision goggles and laser-guided rifles to take out entire herds. Some rent helicopters so they can shoot hogs from above.[94] Others have scattered flavored poison in the hopes that the hogs will take the bait and die.[95] One method, known is the "Judas Pig Technique," is especially fascinating. In this method, biologists capture a feral pig and force the animal to wear a collar with a radio-tracking device. They then set the animal free. Because of the pig's naturally gregarious nature, the animal will invariably seek out others and, in so doing, inadvertently betray the herd's location. When the pigs are located, the entire herd is executed ... save for one. The Judas pig is allowed to survive, free to seek out new herds, unaware of the telemetric albatross around his or her neck. "Once the eradication program has been completed," one biologist explains, "the Judas pig then is located and killed, and the radio collar is retrieved."[96]

Despite all of these efforts and all that ingenuity, feral pigs have proven almost impossible to destroy.[97] One would think recreational hunters might have been able to help. An amateur naturalist suggested as much in *Forest and Stream* as early as 1910. "To curtail the range of these beasts [feral pigs] wherever it may be possible is a line of work that should be taken up by sportsmen's clubs and Audubon societies in every state affected," the correspondent opined.[98] But the interests of biologists and recreational hunters do not overlap as much as one might think. Biologists want to steer ecosystems toward a specific composition they deem healthy. Recreational hunters want to shoot pigs with guns. Those interests may overlap, but they are not the same thing. Many hunters have no interest in getting rid of feral pigs *altogether*. After all, they enjoy shooting the animals, and some of them have economic interests at stake. As early as 1954, visitors to booming post-war Miami paid for the privilege of hunting feral pigs. For $60, a party of four could hire a guide to take them on a feral-pig hunt in the nearby Everglades. Since the execution was accomplished with a pistol at close range, tourists were not allowed to pull the trigger.[99] In other words, people paid money to travel into the wilderness to watch strangers shoot their former commensals in the head at close range.

It is well documented that some hunters have deliberately released pigs into the woods for commercial and recreational purposes.[100] In May, 1962, hunters released twenty-six pen-reared hogs near Crossville, Tennessee, in an attempt to establish "another huntable population," but hunters complained that the animals were too tame.[101] They desired something more. They wanted the thrill of the chase, and they wanted an impressive trophy to showcase after the kill. Pen-reared animals provided neither. To improve (or is it unimprove?) the South's feral hogs, hunters began trapping the animals near Hooper Bald in western North Carolina. The reader will recall from Chapter 5 that Midwestern sportsmen had imported thirteen wild (never-domesticated) boars from Europe, and that hundreds had escaped into the Carolina wilderness by the 1920s. Wild boars are not only much more tenacious than their domesticated cousins, but are also considerably larger. As a result, hunters began trapping these animals and exporting them to numerous other locations throughout the Southeast. By the 1970s, hunters had transported wild boars to sites in Florida, South Carolina, and Georgia, where they freely swapped genes with the region's feral suids.[102]

Owing to these clandestine translocations, a majority of the pigs in the South are now hybrids between wild and feral suids. As a result, the region's free-roaming pigs are now much larger than they were in previous centuries. This development has increased the hunter's desire to kill these animals, and feral pigs are now one of the most popular hunting trophies in North America, second only to whitetail deer in the number killed[103] These developments have also helped invigorate the popular myth that the South is overrun with enormous, tenacious, quasi-mythical swine. Reported sightings of massive hogs have increased dramatically over the past decade. After one image of a giant feral pig went viral in 2004, *National Geographic* sent a team of scientists to southern Georgia to exhume the animal and measure its dimensions. Post-mortem analyses confirmed that the specimen, now known as "Hogzilla," was more than 9 feet long and had probably weighed more than 800 pounds.[104]

An even more dramatic photograph surfaced in 2007. It showed a boy wearing a ball cap, holding a pistol, and kneeling behind what appeared to be an enormous feral pig (Figure 19). The image was reprinted in the nation's major news outlets, and the ten-year-old boy in the picture, Jamison Stone, was interviewed in-studio by CNN. He and his father swore that they had been hunting, came upon the animal (which they dubbed "Monster Pig"), and that young Jamison had felled the beast with eight shots from a .50-calliber revolver. People from around the world cried protest. Some objected to recreational hunting in general, but most people just doubted the veracity of the story altogether. Many claimed that the photograph was an optical illusion, and that no pig could possibly be that large. Others insisted that the photograph had somehow been manipulated.[105] Suspicions were further raised when the boy's father began promoting a website called "monsterpig.com." Well, it turns out the story *was* a fake, but not in the way one might think. About a month after the "Monster Pig" story first broke, reports surfaced that the animal was not, in fact, feral. On the contrary, the animal had been pen-reared for more than ten years on a farm in southern Alabama. The pig's owners, Phil and Rhonda Blissitt, had even given the animal a name: "Fred." Eddy Borden, the owner of Lost Creek Plantation,

Figure 19 Jamison Stone and "Monster Pig," 2007
After this photo went viral in 2007, the boy in the photo, Jamison Stone, was interviewed live in-studio on *CNN*. He and his father claimed that Jamison had killed the feral pig while hunting at Lost Creek Plantation in southern Alabama. It was eventually revealed that "Monster Pig" had been pen-raised for more than ten years, and that his previous owners had even given him a name: "Fred."
**Source** – Brian Stickland, "Monster Pig raised on Fruithurst farm, not a wild hog," *Columbus Ledger-Enquirer*, June 1, 2007.

purchased the animal, turned him loose on a 150-acre enclosure, and advertised the opportunity to hunt a wild boar. By the time Jamison Stone shot Fred to death, the animal had been "feral" for less than a week.[106] Scientists who analyzed the animal's skull confirmed that the pig had indeed been enormous, but that he had also been domesticated.

Edmund Russell has recently shown that hunting can shape a population's genetic composition, and thus its evolutionary fortunes, in profound ways.[107] As an example, he describes how elephants in Africa were hunted for their tusks, and, over the course of several generations, a greater percentage of tuskless elephants roamed the savannas, not because of Lamarckian inheritance, but rather because tuskless elephants had not been previously harvested and their genes thus survived to live another day. When it comes to feral pigs, hunting appears to be driving populations in different directions. Feral-pig hunting has fueled a feral-pig population boom. Even hunters have admitted as much. "There's an undeniable correlation between the wild pig's spread in the past few decades and its glamorization as a big-game animal," *Field & Stream* reported in the summer of 2015, adding that "most new pig problems can be blamed on escaped animals from high-fence operations or illegal stockings by outfitters and hunters to create recreational opportunities."[108] Hunters transplant pigs so they can be closer to their quarry, and thereby extend the incredibly fecund animal's biogeographic distribution. Hunters are far more likely to trap and transplant extant feral hogs and wild boars than they are to deliberately feralize domestic pigs, but they are, in any event, planting the animals so they can kill them.

It should be noted, however, that while biologists generally advocate for the extermination of all feral pigs, they have also singled out one population for exemption. Many insist that the feral pigs who roam undeveloped Ossabaw Island just off the coast of Savannah have acquired unique physiological traits over the past several centuries, and that they should therefore be protected. Genetic tests suggest that the Ossabaw pigs are descended from Spanish pigs, who arrived hundreds of years earlier than their British counterparts.[109] Feral pigs were still roaming the island in the 1770s, when Georgia colonists treated them as a marketable commodity.[110] Reports confirm that feral pigs continued to roam the island for the next 200 years.[111] When ecologists began studying the pig on Ossabaw Island during the 1970s, they learned that centuries of ferality had endowed the animals with unique physiological properties. For example, the pigs, who have long subsisted on little more than marsh plants, can tolerate extremely high concentrations of salt, and scientists are eager to learn how. What is more, the pigs contain the highest lipid reserves of any feral pigs in the world. "This biochemical adaptation is similar to non-insulin dependent diabetes in humans," the American Livestock Breeds Conservancy explains.[112] This makes them a "biomedical treasure" for diabetes researchers.[113] Meanwhile, the Colonial Williamsburg living museum in southeast Virginia began exhibiting Ossabaw pigs in the early 2000s. Since the animals have generally avoided humanity's widespread breeding initiatives for centuries, museum officials explain that they are morphologically similar to colonial pigs.[114] Perhaps, though it should be mentioned that the pigs on Ossabaw have never been entirely isolated from anthropogenic manipulation. People have not only hunted the pigs for hundreds of years, but have also introduced still more pigs over the past hundred years.[115]

Last, but not least, there is one more animal for whom we must account. When we last discussed dogs, their population was cleaving in two. Some dogs were actively encouraged to reproduce. These "purebred" dogs were highly prized and their collective genome fiercely protected. They may have done nothing more than manifest some particular morphology that people found appealing, but that was enough. Other dogs were actively discouraged from reproducing. Known as mutts or curs, these dogs more frequently represented genetic admixtures. No human forethought had deliberately intervened and arranged their respective births, and, as a result, they were increasingly deemed unfit to live. Where do things stand now?

Across the South, millions of people, representing every race, gender and class, choose to cohabitate with dogs, often multiple dogs.[116] Despite this fact, no government has ever seen fit to count the number of dogs in this region (or this nation), and so gauging the size of the South's canine population is difficult.[117] Luckily, the American Veterinary Medical Association (AVMA) began counting the nation's dogs in 1996, and has published new censuses every five years. Between 1996 and 2006, the number of dogs in the five southeastern states rose from about 8,500,000 to 12,500,000, but when the AVMA released data from its most recent census in 2011, there were only 12,000,000 dogs in the region. In other words, the number of dogs had actually *decreased*, even though the number of humans had continued to increase.[118] Once again, the financial meltdown of 2008 appears to have played a primary role in this population contraction.

Southerners have also influenced other aspects of the dogs' evolutionary fortunes, including their biogeographic distribution. As we discussed in the previous chapter, people began migrating to cities during the first half of the twentieth century, but this practice accelerated in dramatic fashion during the second half.[119] "Within fifteen years or so after World War II, the rural South was virtually depopulated," Kirby writes.[120] While none of the southeastern states (save for Florida) were predominantly "urban" in 1950, all of them were at least two-thirds urban in 2010. Florida was more than 90 percent urban.[121] As southerners settle into ever more urbanized (and suburbanized) communities, their dogs' habitats have changed in dramatic fashion. For example, more than half of the nation's dogs now live their entire lives indoors save for brief ventures outside. Nearly three-quarters of dogs sleep inside every night, and, among this number, 45 percent sleep in their human's bed.[122] It may have taken a while, but there is little doubt that the range has finally started to close on dogs as well.

Perhaps most significantly, evidence suggests that domestic dogs are also growing more "refined." According to the AVMA, 54 percent of the nation's pet dogs are "purebred," while the remaining 46 percent are "mixed breed." (The report did not provide state or regional breakdowns.) Let us be clear about what this means. Increasingly, people compel specific dogs to procreate with other specific dogs, so that their progeny will look a certain way. As a result of these efforts, Americans can now choose from among hundreds of carefully cultivated varieties, also known as "breeds."[123] Some of these dogs, like Florida's racing greyhounds, are subject to intense anthropogenic selection in the name of sport, but most of the cultivated dogs serve as companions.[124] Given this intense interest, it should come as no surprise that scientists have already used CRISPR-Cas9 gene-editing technologies to cultivate dogs with extreme morphological features.[125]

The proliferation and variety of canine cultivars would seem to suggest that, from an evolutionary perspective, domestic dogs are flourishing. Many of these animals share a deep, earnest, reciprocal love with their human companions. As well they should, especially since fully two-thirds of dog owners consider their particular canid a "member of the family" rather than a mere "pet."[126] The truth is not that simple, though. It is certainly true that *some* dogs are flourishing, but they represent a narrower subset of the canine genome than most people realize. As Greger Larson explains, "breeders have simultaneously closed breeding lines and selected for extreme morphological traits."[127] In other words, the diverse dogs on this planet do not represent widely disparate branches on the tree of life as much as they represent one particular branch that manifests extreme organismal plasticity. In fact, most dogs can trace their ancestry back to a relatively recent bottleneck in nineteenth-century Europe.[128] Most of the breeds with which we are familiar, including many purportedly "ancient" breeds, were actually cultivated within the past few hundred years or so.[129] Finally, some biologists question whether domestication has truly served the dog's best interests. According to these individuals, domestication has merely facilitated the accumulation of "deleterious" genes that diminish the animal's overall quality of life.[130]

Given how little initiative the United States has shown when it comes to counting domestic dogs, the reader can probably guess how much effort has been directed at counting feral dogs.[131] That began to change in the 1970s, when Americans began keeping much closer tabs on their animals' reproductive output. The impetus may well have been Carl Djerassi's article on the overpopulation of dogs in 1973. Known as one of the inventors of oral contraception for humans ("the pill"), Djerassi called attention to another potential market for chemical contraception. He pointed out that millions of unwanted dogs were euthanized every year, costing public and private shelters tens of millions of dollars.[132] Heeding Djerassi's warnings, many municipalities began to exert greater control over their respective canids. Rather than developing expensive contraceptive pills for dogs, however, most cities employed a mechanical fix. Although castration had been commonly practiced on livestock since time immemorial, the practice was not commonly applied to dogs until the late twentieth century.[133] At the same time, animal-rights advocacy also increased in the region.[134] Finding castration preferable to the existing policy (wholesale eradication), animal-advocacy groups began promoting surgical sterilization as a pet population control method."[135] (Figure 20.)

Over the next few decades, the number of dogs being euthanized decreased dramatically. In the early 1970s, the nation euthanized somewhere between 13,000,000 and 23,000,000 every single year. By the early 1980s, the number had fallen to somewhere between 7,000,000 and 10,000,000 deaths per year. The estimated number of pet euthanasias had dropped to 5,700,000 by 1992 and had dropped to 4,600,000 by the turn of the millennium.[136] Despite these trends, people in the South have not exterminated feral dogs by any measure. Millions still enter shelters annually, and millions still end up being euthanized.[137] That may soon change, though. In 2015, scientists reported that engineered

Figure 20 Photograph of trailer at Miami-Dade Animal Services
Like most communities throughout the South and across the United States, the city of Miami encourages its citizens to remove their pets' reproductive capabilities.
The author took this photograph during a recent visit to Miami-Dade Animal Services.

DNA had rendered laboratory rats sterile, and that the same might prove true for dogs.[138]

To learn more about the state of feral dogs in the South, the author recently queried city officials in the region's largest metropolitan area, Miami. Representatives from Miami-Dade Animal Services generously provided all relevant data and patiently answered all of the author's questions. According to their data, the city's main animal shelter processed 17,734 dogs in 2011. That amounts to 49 new dogs arriving at the shelter every single day. Some of these animals never had a prayer (they were either dead on arrival or euthanasia requests from pet owners). The vast majority of dogs entering the shelter faced a few possible futures, and their fate generally depended on their age. Approximately 54 percent of all puppies were adopted, while just 28 percent of adult dogs were adopted. In similar fashion, 31 percent of adult dogs were euthanized in 2011, compared with "only" 15 percent of puppies. The percentage of deaths would no doubt be much higher, but employees at Miami-Dade Animal Services deserve significant credit for their earnest effort and innovative approaches to minimizing the annual number of euthanasias. In 2011, the city transferred 28 percent of the puppies and 28 percent of the adult dogs to shelters and humane societies across the nation. Officials have made efforts to expand this practice over the past few years, and have expressed willingness to send dogs anywhere there is a demand. Meanwhile, officials state that roughly 180,000 dogs have been registered with the city. They acknowledge that that number represents a fraction of the city's actual canine population, and that as many as 400,000 dogs may live in greater Miami.[139]

The reader will recall that the South has hosted two genetically distinct populations of dogs during its long history. The first population arrived in the region alongside hunter-gatherers around 15,000 years ago, while the second population began arriving with Europeans in the early sixteenth century AD. For years, scholars have maintained that the latter population overwhelmed the former and succeeded in completely replacing it. As early as 1803, Barton wrote that the dogs of southeastern Indians were "mixed," though he retained some hope that a few unadulterated pockets might still exist somewhere on the North American continent. By the early twentieth century, Glover Allen had abandoned that hope. "At the present day it is probably too late to find pure-bred examples of most of the local varieties that formerly occurred," he wrote in 1920.[140] Contemporary scholars generally hold the same opinion. "European dogs rapidly interbred with native dogs and have, for the most part, succeeded in completely replacing them," Marion Schwartz writes, adding that the "American aboriginal dogs" have all but disappeared.[141] Even some scientists have taken this position. In 2011, Castroviejo-Fisher and colleagues conducted genetic analyses that appeared to confirm "the extensive replacement of the Native American dog population."[142]

Yet for the past forty years, one scientist has stubbornly rejected that consensus. Lehr Brisbin, emeritus ecologist at the Savannah River Ecology Lab, first began to question the fate of Native American dogs in the 1970s. One day he was trekking across the Savannah River Site, a vast tract of "wilderness" in western South Carolina that has housed nuclear processing facilities for the past half-century and which doubles as an enormous ecological preserve. He had caught numerous free-ranging dogs while conducting a census of the mammals on the Savannah River Site, though he had generally paid them little mind. They were nearly all the same: medium-sized and yellow-coated with upright ears and a fishhook tail. He had caught and was releasing another one of these ownerless dogs when it suddenly struck him that the animal's appearance was nearly identical to an Australian dingo.[143] The resemblance was extraordinary and, he suspected, more than incidental.[144] For the past several decades, he has stubbornly suggested that some of the free-ranging canids in the Savannah River Valley, creatures he has dubbed "Carolina Dogs," may well be descended from the first dogs who migrated into the region at the end of the Pleistocene.[145] Brisbin has convinced officials at the United Kennel Club, which now lists Carolina Dogs as a distinct breed. Unlike other breeds, however, these dogs have largely evaded intense anthropogenic selection. Unlike other breeds, they owe their morphology, behavior, and genetic composition to natural selection, and not human whims.[146]

Until recently, there were several factors working against Brisbin's hypothesis. For example, the Savannah River Site might be much more lightly trod than the rest of its surroundings, but that has not always been the case. As recently as 1950, the area contained two distinct communities, Ellenton and Dumbarton, with a combined population in excess of a thousand. We even have documented evidence that when all of these people were abruptly uprooted so that the federal government could build nuclear processing facilities in their respective hometowns, some of the displaced people left their dogs behind.[147] It seems entirely possible that these abandoned animals could have been the source of Carolina Dogs, in which case they could be feral dogs of Old-World descent. That being said, there is also evidence in support of Brisbin's hypothesis. For example, zooarchaeologists have discovered the remains of several ancient dogs who were buried on the Savannah River Site long before European contact. More significant still, geneticists recently sequenced several Carolina Dogs and found that their genes were more similar to Asian dogs than to European dogs. They conclude that Carolina Dogs may have "an indigenous American origin and are not just 'run-away' dogs of European descent."[148] Subsequent tests suggest that Carolina Dogs derive some of their genome from dogs who lived in the Americas prior to contact, but most of their genome from dogs who arrived with Europeans.[149] Geneticists insist that further testing will be necessary to clarify the mystery.

Finally, wolves have also carved a new niche in the region. The reader will recall that most of the dog's closest cousins were exterminated from the South by the early nineteenth century. In the late twentieth century, officials with the U.S. Fish and Wildlife Service deemed the "red wolf," an animal whose range purportedly once stretched from Virginia to Texas, endangered. Unable to save the animals from extinction, officials plucked the last few red wolves from the Texas Gulf Coast region in 1980, rendering the animal extinct in the wild. These animals were taken to a breeding center in Washington State for several years, where handlers minimized contact in an effort to reduce socialization and promote "wildness." In 1980, the animals were released in a million-acre wildlife preserve located in eastern North Carolina. Officials later released six wolves into the much more heavily trafficked Great Smoky Mountains National Park, but the experiment did not take and the last surviving wolves were swept up in 1998. Today, about a hundred wolves make their home in eastern North Carolina. Most of these animals were born in the "wild," apart from humans, and most of them now wear radio collars that track their every movement.[150]

*Chapter Summary*

Following the invention of the automobile and its diffusion into the South, the number of domestic horses living in the South plummeted. The population bottomed out in the 1970s, and was actually growing prior to its recent contraction. Most people now living in the Southeast do not see horses on a daily basis. Those who do see horses generally see them from the seat of a passing automobile. Less than 2 percent of southeastern households own a pet horse. In other words, more than 98 percent of people living in the South lack either the resources or the resolve to adopt a horse. Meanwhile, scattered bands of feral horses continue to persist in remote biogeographical islands. Although these populations are technically feral, inasmuch as they roam free, reproduce of their own accord, and avoid humans, they are subject to frequent anthropogenic intervention.

The pig's relationship with humanity has proven more stable, though no less complex. The vast majority of domestic pigs in the South are located in eastern North Carolina, where ever more industrialized total-confinement operations have caused their numbers to skyrocket. Owing to this unique method of husbandry, most people living in the South have not seen a domestic pig in years, and most of the region's domestic pigs live their entire lives, "from birth to the dinner plate," without ever seeing the sky. Meanwhile, the number of feral pigs living in the region has exploded. They now inhabit riverine shallows throughout all five southeastern states, and their numbers continue to grow. Biologists are more convinced than ever that these animals need to be eradicated, and recreational hunters hope to pursue this policy in perpetuity.

Finally, domestic dogs retain their popularity in the modern South, where people of every stripe continue to keep dogs as companions. For most of southeastern history, dogs roamed free and reproduced without direct human intervention, but that has recently changed. For the past hundred years or so, people living in the Southeast have shown increased devotion to their pets. The prime beneficiaries have been a relatively narrow sub-set of the dog genome blessed with extreme morphological plasticity. Dogs are now available in hundreds of different cultivated varieties known as "breeds." Complicating matters, communities throughout the South continue to feralize dogs every single day. Thanks to armies of well-meaning citizens and volunteers, some of these dogs find refuge in the homes of compassionate humans. Despite these heroic efforts, most southeastern communities kill unwanted dogs on a daily basis because there is just no other place to put them. This grim statistic reminds us that much as domestication is an ongoing process, so too is feralization.

# Epilogue

## CULTIVATING FERALITY IN THE ANTHROPOCENE

### LESSONS FOR THE AMERICAN SOUTH AND BEYOND

*You become responsible, forever, for what you have tamed.*
~ Antoine de Saint-Exupéry

A growing number of scientists and scholars believe that Earth has entered a new geological epoch known as the Anthropocene. They explain that human activities have transformed the climate, the continents, and the oceans on a planetary scale, and that humanity's collective influence is now so widespread that it is, quite literally, a global force of nature. Some argue that no place on Earth is immune to anthropogenic influence, and that there are no more truly "wild" places left on the planet.[1] Many researchers would hasten to add that humanity's influence on the biosphere is not only pervasive but also exclusively negative. They point out that humans have destroyed ecosystems around the globe, and that we have ushered in the planet's sixth great extinction event.[2]

The world's feral animals occupy a strange niche in the Anthropocene. The fact that they exist at all is significant, especially since most of the planet's large mammals are rapidly disappearing, but that is not the only thing that makes them conspicuous. After all, feral animals are defined by their purported divorce from anthropogenic selection, and yet they exist on a planet in which every nook and cranny is subject to anthropogenic influence. One would think that feral animals would disappear entirely, that they would, by definition, cease to exist, and yet that is not the case. As the foregoing examples from the American South reveal, feral animals continue to exist because people cannot stop creating them.

Consider the horse. In the colonial era, feralizing horses allowed settlers to focus on their cash crops, but those times are long gone. Nowadays, people feralize horses so that the animals can be romanticized in dramatic settings, and, if future generations want to manufacture the same sublime experience, they will eventually have to feralize additional horses. Strange though it may sound, the feral horses' evolutionary future is actually rather bright when compared to their domestic cousins. Their numbers might be limited, but most feral horses in the South (though admittedly not all) have been granted state or federal protection in perpetuity. This is also true of their more celebrated cousins, the feral mustangs. Humanity's compulsion to romanticize feral horses guarantees they will survive for generations hence. What is that if not evolutionary success?

Meanwhile, people feralize pigs for entirely different reasons. Counterintuitive though it may sound, humanity's desire to hunt and kill our former commensals means that the animals have never been more numerous or more widespread. Some people have even supplemented the feral pigs' gene pool with wild, never-domesticated boars. Well, despite all the interest in southern hogzillas, there is no guarantee that their numbers will continue to grow forever. After all, the percentage of Americans who hunt has steadily dropped for almost forty years.[3] This begs several important questions. What will happen to these animals if people no longer want to kill them? Will we then accede to the advice of biologists, who suggest total eradication? What if companies like Smithfield get their way and people are no longer allowed to see inside pig factories? Would any of us really be okay never seeing *any* pigs, domestic or feral, ever again? What is the end game, and what constitutes evolutionary success?

Finally, people feralize dogs for still different reasons. When scientists first studied the source of feral dogs more than fifty years ago, they concluded that the primary cause of canine feralization was human negligence.[4] This remains one of the largest factors contributing to dog feralization, but many other factors, including job loss, foreclosure, divorce, or health problems, also contribute daily to the feralization of dogs.[5] When dogs are feralized in the South, the animals usually end up at county- or city-run impound facilities. Some are adopted by sympathetic humans. Even so, communities throughout the region continue killing unwanted dogs every single day because there is simply no other place to put them. The evidence suggests that southern communities now euthanize fewer dogs each year than they did forty years ago. One cannot help but wonder: Does fewer feral dogs constitute evolutionary success, and, if so, for whom? How can we serve the best interests of our species' best friends?

Given these diverse experiences, developing a comprehensive strategy to guide humanity's interactions with feral animals in the Anthropocene will not be easy, but it is not impossible, either. We can start by abandoning the idea that feral animals live apart from us. It is clear that humans are once again the primary force shaping the distribution and composition of feral populations. These animals are subject to a much different kind of anthropogenic selection than their still-domesticated cousins, but it is anthropogenic selection nonetheless. In other words, we retain dominion over these animals whether we like it or not. It is a daunting responsibility, but it is also an opportunity to fix what previous generations got wrong. After all, we inherit our forebears' decisions, however noble or shameful they might have been, but that does not mean that we are also beholden to them. We have free will and we can actively change course. The sooner we accept this fact, the sooner we can finally start cultivating a future that affords dignity to the rest of nature, to other animals, and to one another.

# NOTES

### Preface

1 Charles Phineas, "Household Pets and Urban Alienation," *Journal of Social History* 7 (1974): 338–343. Phineas makes such good points with such dry humor that one could easily view his essay as prescient, but most scholars interpret his essay as satire. For examples, see: David Gary Shaw, "A Way with Animals," *History and Theory* 52 (December, 2013): 1–12; Jane Costlow and Amy Nelson, "Foreword," in *Other Animals: Beyond the Human in Russian Culture and History*, edited by Jane Costlow and Amy Nelson (Pittsburgh: University of Pittsburgh, 2010); Erica Fudge, "A Left-Handed Blow: Writing the History of Animals," in *Representing Animals*, edited by Nigel Rothfels (Bloomington: University of Indiana Press, 2002), 4; John K. Walton, "Mad Dogs and Englishmen: The Conflict over Rabies in Late Victorian England," *Journal of Social History* 13 (winter, 1979): 219–239.
2 Harriet Ritvo, "On the Animal Turn," *Daedalus* 136 (fall, 2007): 118–122. There is an entire discipline in the humanities known as "animal studies," but historians remain somewhat marginalized within this field. As Harriet Ritvo recently explained, most practitioners in the field of animal studies prefer to focus on the *idea* of animals, and so their research is characterized by a high degree of "abstractness" and an explicit focus on theory. For more, see: Harriet Ritvo, "Among Animals," *Environment and History* 20 (2014): 491–498. Despite their marginalization within the field of animal studies, historians have produced incredibly sophisticated scholarship on the history of animals. Representative works include, but are certainly not limited to: Brian Fagan, *The Intimate Bond: How Animals Shaped Human History* (New York: Bloomsbury Press, 2015); Susan Nance (ed.), *The Historical Animal* (Syracuse: Syracuse University Press, 2015); Alan Mikhail, *The Animal in Ottoman Egypt* (New York: Oxford University Press, 2014); Sam White, "Animals, Climate Change, and History," *Environmental History* 19 (April, 2014): 319–328; Alan Mikhail, "Unleashing the Beast: Animals, Energy, and the Economy of Labor in Ottoman Egypt," *American Historical Review* (April, 2013): 317–348; Martha Few and Zeb Tortorici (eds.), *Centering Animals in Latin American History* (Durham: Duke University Press, 2013); Etienne Benson, "The Urbanization of the Eastern Gray Squirrel in the United States," *Journal of American History* (2013): 691–710; Susan Nance, *Entertaining Elephants: Animal Agency and the*

*Business of the American Circus* (Baltimore: Johns Hopkins University Press, 2013); Harriet Ritvo, "Going Forth and Multiplying: Animal Acclimatization and Invasion," *Environmental History* 17 (2012): 404–414; Harriet Ritvo, *Noble Cows and Hybrid Zebras: Essays on Animals and History* (Charlottesville: University of Virginia Press, 2010); Alan Mikhail, "Animals as Property in Early Modern Ottoman Egypt," *Journal of the Economic and Social History of the Orient* 53 (2010): 621–652; Dorothee Brantz (ed.), *Beastly Natures: Animals, Humans, and the Study of History* (Charlottesville: University of Virginia Press, 2010); Thomas G. Andrews, "Contemplating Animal Histories: Pedagogy and Politics across Borders," *Radical History Review* 107 (spring, 2010): 139–165; Harriet Ritvo, "Beasts in the Jungle (Or Wherever)," *Daedalus* (2008): 22–30; Linda Kaloff, *Looking at Animals in Human History* (London: Reaktion, 2007); Richard W. Bulliet, *Hunters, Herders, and Hamburgers: The Past and Future of Human–Animal Relationships* (New York: Columbia University Press, 2005); Karen Rader, *Making Mice: Standardizing Animals for American Biomedical Research, 1900–1955* (Princeton: Princeton University Press, 2004); Harriet Ritvo, "Animal Planet," *Environmental History* 9 (April, 2004): 204–220; Nigel Rothfels, *Savages and Beasts: The Birth of the Modern Zoo* (Baltimore: Johns Hopkins University Press, 2002); Louise Robbins, *Elephant Slaves and Pampered Parrots: Exotic Animals in Eighteenth-Century Paris* (Baltimore: Johns Hopkins University Press, 2002); Harriet Ritvo, "History and Animal Studies," *Society & Animals* 10 (2002): 403–406; Harriet Ritvo, "Animal Consciousness: Some Historical Perspective," *American Zoologist* (2000): 847–852; Kathleen Kete, *The Beast in the Boudoir: Petkeeping in Nineteenth-Century Paris* (Berkeley: University of California Press, 1994); Harriet Ritvo, *The Animal Estate: The English and Other Creatures in the Victorian Age* (Cambridge: Harvard University Press, 1987). To learn about the field of animal studies, see: Margo DeMello, *Animals and Society: An Introduction to Human–Animal Studies* (New York: Columbia University Press, 2012); Aaron Gross and Anne Vallely, *Animals and the Human Imagination: A Companion to Animal Studies* (New York: Columbia University Press, 2012); Arnold Arluke and Clinton Sanders (eds.), *Between the Species: Readings in Human–Animal Relations* (Boston: Pearson, 2009); Clifton P. Flynn (ed.), *Social Creatures: A Human and Animal Studies Reader* (New York: Lantern, 2008); Cary Wolfe, *What Is Posthumanism?* (Minneapolis: University of Minnesota Press, 2010); Donna Haraway, *When Species Meet* (Minneapolis: University of Minnesota Press, 2008); Cary Wolfe, *Animal Rites: American Culture, the Discourse of Species, and Posthumanist Theory* (Chicago: University of Chicago Press, 2003); Donna Haraway, *The Companion Species Manifesto: Dogs, People, and Significant Otherness* (Chicago: University of Chicago Press, 2003); Erica Fudge, *Perceiving Animals: Humans and Beasts in Early Modern English Culture* (New York: Basingstoke, 2000); Carol J. Adams, *The Sexual Politics of Meat: A Feminist-Vegetarian Critical Theory* (New York: Continuum, 2000); Steve Baker, *Picturing the Beast: Animals, Identity, and Representation* (Manchester: Manchester University Press, 1993); Keith Thomas, *Man and the Natural World: Changing Attitudes in England, 1500–1800* (London: Allen Lane, 1983); John Berger, "Why Look at Animals?" in *About Looking*, edited by John Berger (New York: Pantheon, 1980), 3–28.

3 There is a lively debate surrounding the use of "who" when referring to non-human animals. For more on the topic of animals and agency, see: Paul S. Sutter, "The World with Us: The State of American Environmental History," *Journal of American History* (June, 2013): 94–119; Chris Pearson, "Dogs, History, and Agency," *History and Theory* 52 (December, 2013): 128–145.

## 1 The Trouble with Ferality

1 Melinda A. Zeder, "The Domestication of Animals," *Journal of Anthropological Research* 68 (summer, 2012): 170–171; Helmut Hemmer, *Domestication: the Decline of Environmental Appreciation*, translated into English by Neil Beckhaus (Cambridge: Cambridge University Press, 1990), 159–160.
2 Georges Louis Leclerc Buffon, *Natural History*, vol. 2 (London: Proprietor, 1797), 38.
3 Thomas Jefferson, *Notes on the State of Virginia* (London: John Stockdale, 1787), 90. See also: Lee Alan Dugatkin, *Mr. Jefferson and the Giant Moose: Natural History in Early America* (Chicago: University of Chicago Press, 2009).
4 Charles Darwin, *Variation of Animals and Plants under Domestication*, vol. 1 (London: John Murray, 1868), 27–28, 51–52, 77–78, 112–130; Charles Darwin, *Variation of Animals and Plants under Domestication*, vol. 2 (London: John Murray, 1868), 32.
5 Alfred Russel Wallace, "On the Tendency of Varieties to Depart Indefinitely from the Original Type," *Contributions to the Theory of Natural Selection* (London: Macmillan and Company, 1870), 33, 41.
6 *The Thematic Atlas of Project Isabela: An Illustrative Document Describing, Step-by-Step, the Biggest Successful Goat Eradication Project on the Galapagos, 1998–2006* (Charles Darwin Foundation / Galapagos National Park Service, 2007); Tim Low, *Feral Future: The Untold Story of Australia's Exotic Invaders* (Chicago: University of Chicago Press, 1999).
7 Tom McKnight, *Feral Livestock in Anglo-America* (Berkeley: University of California Press, 1964), 3.
8 Thomas J. Daniels and Marc Bekoff, "Feralization: The Making of Wild Domestic Animals," *Behavioural Processes* 19 (1989): 80.
9 I. Lehr Brisbin, "The Pariah: Its Ecology and Importance to the Origin, Development and Study of Pure Bred Dogs," *American Kennel Gazette* 94 (January, 1977): 22–23. See also: Daniels and Bekoff, "Feralization," 87–88.
10 Jay Kirkpatrick, *Into the Wind: Wild Horses of North America* (Minocqua, WI: Northwood Press, 1994), 23–25.
11 Jonaki Bhattacharyya and D. Scott Slocombe, "The 'Wild' or 'Feral' Distraction: Effects of Cultural Understandings on Management Controversy over Free-Ranging Horses," *Human Ecology* 39 (2011): 613–625.
12 Alfred Crosby, *Ecological Imperialism: The Biological Expansion of Europe* (New York: Cambridge University Press, 1986); Alfred Crosby, *The Columbian Exchange: Biological and Cultural Consequences of 1492* (Westport: Greenwood, 1972).
13 Elinor G. K. Melville, *A Plague of Sheep: Environmental Consequences of the Conquest of Mexico* (Cambridge: Cambridge University Press, 1997).

14 Virginia DeJohn Anderson, *Creatures of Empire: How Domestic Animals Transformed Early America* (Oxford: Oxford University Press, 2004).
15 Harriet Ritvo, "Race, Breed, and Myths of Origin: Chillingham Cattle as Ancient Britons," *Representations* 39 (summer, 1992): 1–22.
16 Many writers study feral animals within the context of commensal and/or invasive species. For examples, see: Tristan Donovan, *Feral Cities: Adventures with Animals in the Urban Jungle* (Chicago: Chicago Review Press, 2015); Kelsi Nagy and Phillip David Johnson II (Eds.), *Trash Animals: How We Live with Nature's Filthy, Feral, Invasive, and Unwanted Species* (Minneapolis: University of Minnesota Press, 2013); Terry O'Connor, *Animals as Neighbors: The Past and Present of Commensal Species* (Lansing: Michigan State University Press, 2013). Others examine feral animals within the context of "rewilding." For example, see: George Monbiot, *Feral: Search for Enchantment on the Frontiers of Rewilding* (Harmondsworth: Penguin Books, 2013).
17 Michael Pollan, *Botany of Desire: A Plant's-Eye View of the World* (New York: Random House, 2001).
18 Edmund Russell, *Gambling on Evolution: the Coevolution of English Culture with Fierce, Fleet, and Fancy Dogs* (New York: Cambridge University Press, forthcoming); Edmund Russell, "Coevolutionary History," *American Historical Review* (December, 2014): 1514–1528; Edmund Russell, *Evolutionary History: Uniting History and Biology to Understand Life on Earth* (New York: Cambridge University Press, 2011); Edmund Russell, "Evolutionary History: Prospectus for a New Field," *Environmental History* 8 (April, 2003): 204–228. See also: Margaret Derry, *Art and Science in Breeding: Creating Better Chickens* (Toronto: University of Toronto Press, 2012); Sam White, "From Capitalist Pigs to Globalized Pig Breeds: A Study in Animal Cultures and Evolutionary History," *Environmental History* 16 (January, 2011): 94–120; Ann Norton Greene, *Horses at Work: Harnessing Power in Industrial America* (Cambridge: Harvard University Press, 2008); Margaret E. Derry, *Bred for Perfection: Shorthorn Cattle, Collies, and Arabian Horses since 1800* (Baltimore: Johns Hopkins University Press, 2003); Susan Schrepfer and Philip Scranton (Eds.), *Industrializing Organisms: Introducing Evolutionary History* (New York: Routledge, 2003). To learn more about collaborations between biologists and historians, see: Julia Adeney Thomas, "History and Biology in the Anthropocene: Problems of Scale, Problem of Value," *American Historical Review* (December, 2014): 1587–1607.
19 Roderick Nash, *Wilderness and the American Mind* (New Haven: Yale University Press, 1967). See also: Max Oelschlaeger, *The Idea of Wilderness* (New Haven: Yale University Press, 1991).
20 William Cronon, "The Trouble with Wilderness: Or, Getting Back to the Wrong Nature," *Environmental History* 1 (January, 1996): 7–28.
21 Paul S. Sutter, "The World with Us: The State of American Environmental History," *Journal of American History* (June, 2013): 94–119. The classic texts on hybridity remain: Richard White, *Organic Machine: The Remaking of the Columbia River* (New York: Hill and Wang, 1996); Michael Pollan, *Second Nature: A Gardener's Education* (New York: Delta, 1992); William Cronon, *Nature's Metropolis: Chicago and the Great West* (New York: Norton and Company, 1991). For more on the slippery nature of wilderness, see: Etienne Benson,

"From Wild Lives to Wildlife and Back," *Environmental History* 16 (2011): 418–422; Etienne Benson, *Wired Wilderness: Technologies of Tracking and the Making of Modern Wildlife* (Baltimore: Johns Hopkins University Press, 2010); Michael P. Nelson and J. Baird Callicott (eds.), *The Wilderness Debate Rages On: Continuing the Great New Wilderness Debate* (Athens: University of Georgia Press, 2008); Michael Lewis (ed.), *American Wilderness: A New History* (New York: Oxford University Press, 2007); Rebecca Cassidy and Molly Mullin, *Where the Wild Things Are Now: Domestication Reconsidered* (New York: Berg Publishers, 2007); John Wills, "'On Burro'd Time': Feral Burros, the Brighty Legend, and the Pursuit of Wilderness in the Grand Canyon," *Journal of Arizona History* 44 (2003): 1–24; J. Baird Callicott and Michael P. Nelson (eds.), *The Great New Wilderness Debate* (Athens: University of Georgia Press, 1998).

22 The literature on dogs is enormous. For works that examine the history of dogs in general, see: Laura Hobgood-Oster, *A Dog's History of the World: Canines and the Domestication of Humans* (Waco: Baylor University Press, 2014); Mark Derr, *How the Dog Became the Dog: From Wolves to Our Best Friends* (London: Overlook, 2011); Jon Franklin, *The Wolf in the Parlor: How the Dog Came to Share Your Brain* (New York: Macmillan, 2009); Mark Derr, *A Dog's History of America: How Our Best Friend Explored, Conquered, and Settled a Continent* (New York: North Point Press, 2004); Stanley Coren, *The Pawprints of History: Dogs and the Course of Human Events* (New York: Simon and Schuster, 2002); Raymond Coppinger and Lorna Coppinger, *Dogs: A Startling New Understanding of Canine Origin, Behavior, and Evolution* (New York: Scribner, 2001). For works that examine dogs in specific regions and/or time periods, see: Philip Howell, "The Dog Fancy at War: Breeds, Breeding, and Britishness, 1914–1918," *Society & Animals* 21 (2013): 546–567; Lance van Sittert and Sandra Swart (eds.), *Canis Africanis: A Dog History of Southern Africa* (Leiden: Brill, 2008); Jessica Wang, "Dogs and the Making of the American State: Voluntary Association, State Power, and the Politics of Animal Control in New York City, 1850–1920," *Journal of American History* 98 (March, 2012): 998–1024; James Boyce, "Canine Revolution: The Social and Environmental Impact of the Introduction to Tasmania," *Environmental History* 11 (January, 2006): 102–129; Aaron Herald Skabelund, *Empire of Dogs: Canines, Japan, and the Making of the Modern Imperial World* (Ithaca: Cornell University Press, 2011). Several works examine the ecology of feral dogs. For examples, see: Matthew E. Gompper, *Free-Ranging Dogs and Wildlife Conservation* (New York: Oxford University Press, 2014); Stephen Spotte, *Societies of Wolves and Free-Ranging Dogs* (New York: Cambridge University Press, 2012); John L. Long, "Feral Dog," in *Introduced Mammals of the World: Their History, Distribution, and Influence*, edited by John L. Long (Collingwood, Australia: CSIRO, 2003), 249–256; Alan M. Beck, *The Ecology of Stray Dogs: A Study of Free-Ranging Urban Animals* (Indianapolis: Purdue University Press, 1973); M. Douglass Scott and Keith Causey, "Ecology of Feral Dogs in Alabama," *Journal of Wildlife Management* 37 (1973): 253–265.

23 Regarding the history of pigs in general, see: Barry Estabrook, *Pig Tales: An Omnivore's Quest for Sustainable Meat* (New York: Norton and Co, 2015); Mark Essig, *Lesser Beasts: A Snout-to-Tail History of the Humble Pig* (New York: Basic Books, 2015); Brett Mizelle, *Pig* (London: Reaktion, 2011). On pigs in different

regions and/or time periods, see: Ted Genoways, *The Chain: Farm, Factory, and the Fate of Our Food* (New York: Harper, 2014); Catherine McNeur, "The 'Swinish Multitude': Controversies over Hogs in Antebellum New York City," *Journal of Urban History* 37 (2011): 639–660; Sam White, "From Globalized Pig Breeds to Capitalist Pigs: A Study in Animal Cultures and Evolutionary History," *Environmental History* 16 (January, 2011): 94–120; Mark R. Finlay, "Hogs, Antibiotics, and the Industrial Environments of Postwar Agriculture," in *Industrializing Organisms: Introducing Evolutionary History*, edited by Susan R. Schrepfer and Philip Scranton (New York: Routledge: 2004), 237–260. Regarding feral pigs, see: Jeffrey Greene, *The Golden-Bristled Boar: Last Ferocious Beast of the Forest* (Charlottesville: University of Virginia Press, 2012); Ian Frazier, "Hogs Wild," *New Yorker*, December 12, 2005, 71–83; John J. Mayer and I. Lehr Brisbin (eds.), *Wild Pigs: Biology, Damage, Control Techniques and Management* (Aiken: Savannah River National Laboratory, 2009).

24 For works that examine the history of horses in general, see: Wendy Williams, *The Horse: The Epic History of Our Noble Companion* (New York: Scientific American, 2015); Pita Kelekna, *The Horse in Human History* (Cambridge and New York: Cambridge University Press, 2009); Elaine Walker, *Horse* (London: Reaktion, 2008); D.S. Mills, *The Domestic Horse: The Origins, Development and Management of its Behaviour* (Cambridge and New York: Cambridge University Press, 2005); Sandra Olsen (ed.), *Horses through Time* (Lanham: Rowman and Littlefield, 2003); Juliett Clutton-Brock, *Horse Power: A History of the Horse and Donkey in Human Societies* (Cambridge: Harvard University Press, 2002). Works that examine horses in specific regions and/or time periods include: Katherine C. Mooney, *Race Horse Men: How Slavery and Freedom Were Made at the Racetrack* (Cambridge: Harvard University Press, 2014); Sandra Swart, *Riding High: Horses, Humans and History in South Africa* (Johannesburg: Wits University Press, 2010); Greene, *Horses at Work*; Pekka Hämäläinen, *The Comanche Empire* (New Haven: Yale University Press, 2008); Donna Landry, *Noble Brutes: How Eastern Horses Transformed English Culture* (Baltimore: Johns Hopkins University Press, 2008); Clay McShane and Joel A. Tarr, *The Horse in the City: Living Machines in the Nineteenth Century* (Baltimore: Johns Hopkins University Press, 2007); Pekka Hämäläinen, "The Rise and Fall of Plains Indian Horse Cultures," *Journal of American History* 90 (December, 2003): 833–862; Greg Bankoff and Sandra Swart (eds.), *Breeds of Empire: The 'Invention' of the Horse in Southeast Asia and Southern Africa, 1500–1950* (Copenhagen: NIAS, 2007); Margaret E. Derry, *Horses in Society: A Story of Animal Breeding and Marketing Culture, 1800–1920* (Toronto: University of Toronto Press, 2006). Works that examine feral horses include: Bonnie Urquhart Gruenberg, *The Wild Horse Dilemma: Conflicts and Controversies of the Atlantic Coast Herds* (Strasburg, PA: Quagga, 2015); J. Edward de Steiguer, *Wild Horses of the West: History and Politics of America's Mustangs* (Tucson: University of Arizona Press, 2011); Julie Campbell, *The Horse in Virginia: An Illustrated History* (Charlottesville: University of Virginia Press, 2010); John S. Hockensmith, *Spanish Mustangs in the Great American West: Return of the Horse* (Norman: University of Oklahoma Press, 2009); Deanne Stillman, *Mustang: The Saga*

*of the Wild Horse in the American West* (New York: Mariner, 2009); Carmine Prioli and Scott Taylor, *The Wild Horses of Shackleford Banks* (Winston Salem: John F. Blair Publishing, 2007); Bonnie Urquhart Gruenberg, *Hoofprints in the Sand: Wild Horses of the Atlantic Coast* (Strasburg: Eclipse Press, 2002).
25 Regarding differential rates of animal domestication among the various peoples of Earth: Jared Diamond, *Guns, Germs, and Steel: The Fates of Human Societies* (New York: Norton, 1999).
26 The literature on domestication is enormous. For the most recent developments, see: Richard C. Francis, *Domesticated: Evolution in a Man-Made World* (New York: Norton, 2015); Pascale Gerbault, et al., "Storytelling and Story Testing in Domestication," *Proceedings of the National Academy of Sciences* 111 (April 29, 2015): 6159–6154; Melinda Zeder, "Core Questions in Domestication Research," *Proceedings of the National Academy of Sciences* 112 (March 17, 2015): 3191–3198; Greger Larson, et al., "Current Perspectives and the Future of Domestication Studies," *Proceedings of the National Academy of Sciences* 111 (April 29, 2014): 6139–6146; Fiona B. Marshall, et al., "Evaluating the roles of directed breeding and gene flow in animal domestication," *Proceedings of the National Academy of Sciences* 111 (April 29, 2014): 6153–6158; Kristen J. Gremilliona, et al., "Particularlism and the Retreat from Theory in the Archaeology of Agricultural Origins," *Proceedings of the National Academy of Sciences* 111 (April 29, 2014): 6171–6177; Margo DeMello, "The Present and Future of Animal Domestication," in *A Cultural History of Animals in the Modern Age*, edited by Randy Malamud (New York: Bloomsbury, 2011), 67–94.
27 William F. Ruddiman, et al., "Defining the Epoch We Live In," *Science* 348 (April 3, 2015): 38–39; Simon L. Lewis and Mark A. Maslin, "Defining the Anthropocene," *Nature* 519 (March 12, 2015): 171–180; Richard Monastersky, "The Human Age," *Nature* 519 (March 12, 2015): 144–147; Jonathan Williams and Paul J. Crutzen, "Perspectives on Our Planet in the Anthropocene," *Environmental Chemistry* 10 (2013): 269–280; David Oldroyd and Robert Davis, "Inventing the Present: Historical Roots of the Anthropocene," *Earth Sciences History* 30 (2011): 63–84.; Will Steffen, Jacques Grinevald, Paul Crutzen, and John McNeill, "The Anthropocene: Conceptual and Historical Perspectives," *Philosophical Transactions of the Royal Society A* (January, 2011): 842–867; Will Steffen, et al., "The Anthropocene: From Global Change to Planetary Stewardship," *Ambio* 40 (2011): 739–761; Paul J. Crutzen and Will Steffen, "How Long Have We Been in the Anthropocene Era?" *Climatic Change* 61 (2003): 251–257. Some regard the Anthropocene and the Holocene as one and the same. See: Giacomo Certini and Riccardo Scalengh, "Holocene as Anthropocene," *Science* 349 (July 17, 2015): 246.
28 Writing a book requires tough editorial decisions. To learn about some of the feral populations in the South that we will *not* be covering, see: Pamela Jo Hatley, "Feral Cat Colonies in Florida: The Fur and Feathers are Flying," *Journal of Land Use & Environmental Law* 18 (Spring, 2013): 441–465; Kerrie Anne T. Lloyd and Sonia M. Hernandez, "Public Perceptions of Domestic Cats and Preferences for Feral Cat Management in the Southeastern United States," *Anthrozoos* 25 (September, 2012): 337–351; Michael E. Dorcas and

John D. Willson, *Invasive Pythons in the United States: Ecology of an Introduced Predator* (Athens: University of Georgia Press, 2011); Claire Strom, *Making Catfish Bait Out of Government Boys: The Fight against Cattle Ticks and the Transformation of the Yeoman South* (Athens: University of Georgia Press, 2009); Mart A. Stewart, "'Whether Wast, Deodand, or Stray': Cattle, Culture, and the Environment in Early Georgia," *Agricultural History* 65 (1991): 1–29; John Solomon Otto, "Livestock-Raising in Early South Carolina, 1670–1700: Prelude to the Rice Plantation Economy," *Agricultural History* 61 (autumn, 1987): 13–24; W. Theodore Mealor, Jr. and Merle C. Prunty, "Open-Range Ranching in Southern Florida," *Annals of the Association of American Geographers* 66 (September, 1976): 360–376.

29 Michelle Nickerson and Darren Dochuk, *Sunbelt Rising: The Politics of Space, Place, and Region* (Philadelphia: University of Pennsylvania Press, 2013); Juliana Barr, "How Do You Get from Jamestown to Santa Fe?: A Colonial Sun Belt," *Journal of Southern History* 73 (August, 2007): 553–566; William C. Foster, *Climate and Culture Change in North America, AD 900–1600* (Austin: University of Texas Press, 2004).

30 Regarding southern exceptionalism, see: William A. Link, *Southern Crucible: The Making of an American Region* (New York: Oxford University Press, 2015); Christopher A. Cooper, "Rethinking the Boundaries of the South," *Southern Cultures* 16 (2010): 72–88; Orville Vernon Burton, "The South as 'Other,' the Southerner as 'Stranger,'" *Journal of Southern History* (2013): 7–50; Laura F. Edwards, *The People and their Place: Legal Culture and the Transformation of Inequality in the Post-Revolutionary South* (Chapel Hill: University of North Carolina Press, 2009); W.J. Cash, *The Mind of the South* (New York: Knopf, 1941); Twelve Southerners, *I'll Take My Stand: The South and the Agrarian Tradition* (New York: Harper and Brothers, 1930). Works that question southern exceptionalism include: Matthew D. Lassiter and Joseph Crespino, "Introduction: The End of Southern History," in *The Myth of Southern Exceptionalism*, edited by Mattew D. Lassiter and Joseph Crespino (New York: Oxford University Press, 2010), 3–22; Matthew D. Lassiter and Kevin M. Kruse, "The Bulldozer Revolution: Suburbs and Southern History since World War II," *Journal of Southern History* 75 (August, 2009): 691–706; Karen L. Cox, "The South and Mass Culture," *Journal of Southern History* 75 (August, 2009): 677–690; Peter Kolchin, "The South and the World," *Journal of Southern History* 75 (August, 2009): 565–580; Laura F. Edwards, "Southern History as U.S. History," *Journal of Southern History* 75 (August, 2009): 533–564; Byron E. Shafer and Richard Johnston, *The End of Southern Exceptionalism: Class, Race, and Partisan Change in the Postwar South* (Cambridge: Harvard University Press, 2006); Howard W. Odum, *Southern Regions of the United States* (Chapel Hill: University of North Carolina Press, 1936), 5.

31 Paul Sutter (ed.), *Environmental History and the American South: A Reader* (Athens: University of Georgia Press, 2009); Christopher Morris, "A More Southern Environmental History," *Journal of Southern History* 75 (August, 2009): 581–598; Mart A. Stewart, "Southern Environmental History," in *A Companion to the American South*, edited by John B. Boles (Hoboken: Wiley,

2008), 409–423; Martin Melosi (ed.), *The New Encyclopedia of Southern Culture, Volume 8: Environment* (Chapel Hill: University of North Carolina Press, 2007); Jack Temple Kirby, *Mockingbird Song: Ecological Landscapes of the South* (Chapel Hill: University of North Carolina Press, 2006); Donald Davis (ed.), *Southern United States: An Environmental History* (Santa Barbara: ABC-CLIO, 2006); Mart Stewart, "If John Muir had been an Agrarian: American Environmental History West and South," *Environment and History* 11 (2005): 139–162; Otis L. Graham, "Again the Backward Region? Environmental History in and of the American South," *Southern Cultures* 6 (summer, 2000): 50–72; Albert Cowdrey, *This Land, This South: An Environmental History* (Lexington: University of Kentucky Press, 1996).

32 Walter Johnson, *River of Dark Dreams: Slavery and Empire in the Cotton Kingdom* (Cambridge: Harvard University Press, 2013); Christopher Morris, *The Big Muddy: An Environmental History of the Mississippi and Its People from Hernando De Soto to Hurricane Katrina* (New York: Oxford University Press, 2012); Mikko Saikku, *This Delta, This Land: An Environmental History of the Yazoo-Mississippi Floodplain* (Athens: University of Georgia Press, 2011); François Furstenberg, "The Significance of the Trans-Appalachian Frontier in Atlantic History," *American Historical Review* 113 (June, 2008): 648; James W. Byrkit, "Land, Sky, and People: The Southwest Defined," *Journal of the Southwest* 34 (autumn, 1992): 266; James C. Cobb, *The Most Southern Place on Earth: The Mississippi Delta and the Roots of Southern Identity* (New York: Oxford University Press, 1992); Malcolm J. Rohrbough, *Trans-Appalachian Frontier: People, Societies, and Institutions, 1775–1850* (New York: Oxford University Press, 1978).

33 Admittedly, parts of western Florida and western Georgia fall within the Gulf watershed.

34 Drew A. Swanson, "Endangered Species and Threatened Landscapes in Appalachia: Managing the Wild and the Human in the American Mountain South," *Environment and History* 18 (February, 2012): 35–60; Donald E. Davis, *Where There Are Mountains: An Environmental History of the Southern Appalachians* (Athens: University of Georgia Press, 2011); Kathryn Newfront, *Blue Ridge Commons: Environmental Activism and Forest History in Western North Carolina* (Athens: University of Georgia Press, 2011); Sara M. Gregg, *Managing the Mountains: Land Use Planning, the New Deal, and the Creation of a Federal Landscape in Appalachia* (New Haven: Yale University Press, 2010); Timothy Silver, *Mount Mitchell and the Black Mountains: An Environmental History of the Highest Peaks in Eastern North America* (Chapel Hill: University of North Carolina Press, 2003); Chad Montrie, *To Save the Land and People: A History of Opposition to Surface Coal Mining in Appalachia* (Chapel Hill: University of North Carolina Press, 2003); Margaret Lynn Brown, *The Wild East: A Biography of the Great Smoky Mountains* (Gainesville: University of Florida Press, 2001).

35 Paul S. Sutter, *Let Us Now Praise Famous Gullies: Providence Canyon and the Soils of the South* (Athens: University of Georgia Press, 2015); Drew A. Swanson, *A Golden Weed: Tobacco and Environment in the Piedmont South* (New Haven: Yale University Press, 2014); Evan P. Bennett, *When Tobacco Was King: Families, Farm Labor, and Federal Policy in the Piedmont* (Gainesville: University Press of Florida, 2014); Lynn A. Nelson, *Pharsalia: An Environmental Biography of a*

*Southern Plantation, 1780–1880* (Athens: University of Georgia Press, 2010); Paul S. Sutter, "What Gullies Mean: Georgia's 'Little Grand Canyon' and Southern Environmental History," *Journal of Southern History* 76 (August, 2010): 579–598; Benjamin R. Cohen, "Surveying Nature: Environmental Dimensions of Virginia's First Scientific Survey, 1835–1842," *Environmental History* 11 (January, 2006): 37–69; Matthew A. Lockhart, "'The Trouble with Wilderness' Education in the National Park Service: The Case of the Lost Cattle Mounts of Congaree," *Public Historian* 28 (spring, 2006): 11–30; Joyce E. Chaplin, *An Anxious Pursuit: Agricultural Innovation and Modernity in the Lower South, 1730–1815* (Chapel Hill: University of North Carolina Press, 1996); Stanley W. Trimble, *Man-Induced Erosion on the Southern Piedmont, 1700–1970* (Ankeny, IA: Soil and Water Conservation Society, 2008 [1974]).

36 Drew A. Swanson, *Remaking Wormsloe Plantation: The Environmental History of a Lowcountry Landscape* (Athens: University of Georgia Press, 2012); Roy T. Sawyer, *America's Wetland: An Environmental and Cultural History of Tidewater Virginia and North Carolina* (Charlottesville: University of Virginia Press, 2010); Max Edelson, *Plantation Enterprise in Colonial South Carolina* (Cambridge: Harvard University Press, 2009); Megan Kate Nelson, *Trembling Earth: A Cultural History of the Okefenokee Swamp* (Athens: University of Georgia Press, 2005); Mart Stewart, *What Nature Suffers to Groe: Life, Labor, and Landscape on the Georgia Coast, 1680–1920* (Athens: University of Georgia Press, 2002); David McCally, *The Everglades: An Environmental History* (Gainesville: University of Florida Press, 2000); James J. Miller, *An Environmental History of Northeast Florida* (Gainesville: University Press of Florida, 1998); Jack Temple Kirby, *Poquosin: A Study of Rural Landscape and Society* (Chapel Hill: University of North Carolina Press, 1995).

37 Regarding the climate's role in unifying the South and rendering it distinct, see: A. Cash Koeniger, "Climate and Southern Distinctiveness," *Journal of Southern History* 54 (1988): 21–44; Carl N. Degler, *Place over Time: The Continuity of Southern Distinctiveness* (1977); Clarence Cason, *Ninety Degrees in the Shade* (Chapel Hill: University of North Carolina Press, 1935); U.B. Phillips, *Life and Labor in the Old South* (Boston: Little, Brown and Company, 1929). Increasingly, scholars recognize that the climate has indeed influenced history, but that its influence has been far more dynamic and far more heterogeneous than previous generations appreciated. For examples, see: Samuel White, Kenneth M. Sylvester, and Richard Tucker, "North American Climate History," in *Cultural Dynamics of Climate Change and the Environment in North America*, edited by Bernd Sommer (Leiden: Brill, 2015), 109–138; Sam White, "'Shewing the difference betweene their conjuration, and our invocation on the name of God for rayne': Weather, Prayer, and Magic in Early American Encounters," *William and Mary Quarterly* 72 (January, 2015): 33–56; Dennis Blanton, "The Factors of Climate and Weather in Sixteenth-Century La Florida," in *Native and Spanish New Worlds: Sixteenth-Century Entradas in the American Southwest and Southeast*, edited by Clay Mathers, Jeffrey M. Mitchem, and Charles M. Haecker (Tuscon: University of Arizona Press, 2013), 99–121; Cary J. Mock, et al., "The Winter of 1827–1828 over Eastern North America: A Season of Extraordinary Climatic Anomalies,

Societal Impacts, and False Spring," *Climatic Change* 83 (2007): 87–115; Timothy Silver, "Learning to Live with Nature: Colonial Historians and the Southern Environment," *Journal of Southern History* 73 (August, 2007): 539–552; Foster, *Climate and Culture Change in North America*; Mart A. Stewart, "'Let us Begin with the Weather?': Climate, Race, and Cultural Distinctiveness in the American South," in *Nature and Society in Historical Context*, edited by M. Teich et al. (Cambridge: Cambridge University Press, 1996), 240–256; Joyce E. Chaplin, "Climate and Southern Pessimism: The Natural History of an Idea, 1500–1800," in *The South as an American Problem*, edited by Larry J. Griffin and Don Harrison Doyle (Athens: University of Georgia Press, 1995), 57–82; Karen Ordahl Kupperman, "The Puzzle of the American Climate in the Early Colonial Period," *American Historical Review* 87 (December, 1982): 1262–1289; Karen Ordahl Kupperman, "Fear of Hot Climates in the Anglo-American Colonial Experience," *William and Mary Quarterly* 41 (April, 1984): 213–240; C. Vann Woodward, *The Burden of Southern History* (Baton Rouge: Louisiana State University Press, 1960).

38 The United States census refers to this region (plus West Virginia, Maryland, and Delaware) as the "South Atlantic." Timothy Silver also uses this term in his excellent book, *New Face on the Countryside: Indians, Colonists, and Slaves in South Atlantic Forests, 1500–1800* (Cambridge University Press, 1990), which covers the land from Virginia to Georgia (but not Florida). That being said, the "South Atlantic" label has always made me think of the Falkland Islands and its surroundings. Regarding the claim that there are many souths, see: Christopher A. Cooper and H. Gibbs, "Rethinking the Boundaries of the South," *Southern Culture* 16 (winter, 2010): 72–88; Ira Berlin, *Generations of Captivity: A History of African-American Slaves* (Cambridge: Harvard University Press, 2009); Peter Kolchin, *A Sphinx on the American Land: The Nineteenth-Century South in Comparative Perspective* (Baton Rouge: Louisiana State University Press, 2003). See also: J. William Harris, *Deep Souths: Delta, Piedmont, and the Sea Island Society in the Age of Segregation* (Baltimore: Johns Hopkins University Press, 2001); Phillips, *Life and Labor in the Old South*; Edward Ayers, *The Promise of the New South* (New York: Oxford University Press, 1992).

39 Silver, "Learning to Live with Nature," 539–552; Silver, *New Face on the Countryside*; Stewart, *What Nature Suffers to Groe*; Anderson, *Creatures of Empire*; Edelson, *Plantation Enterprise in Colonial South Carolina*; Joyce E. Chaplin, *Subject Matter: Technology, the Body, and Science on the Anglo-American Frontier, 1500–1676* (Cambridge: Harvard University Press, 2001).

40 Jack P. Greene, "Early Modern Southeastern North America and the Broader Atlantic and American Worlds," *Journal of Southern History* 73 (August, 2007): 530–531; April Lee Hatfield, "Colonial Southeastern Indian History," *Journal of Southern History* 73 (August, 2007): 567–578; Greg O'Brien, "The Conqueror Meets the Unconquered: Negotiating Cultural Boundaries on the Post-Revolutionary Southern Frontier," *Journal of Southern History* 67 (February, 2001): 39–72; John Solomon Otto, *The Southern Frontiers, 1607–1860: The Agricultural Evolution of the Colonial and Antebellum South* (Westport: Greenwood, 1989); W. Stitt Robinson, *The Southern Colonial Frontier, 1607–1763* (Albuquerque: University of New Mexico Press, 1979);

Avery Craven, "Turner Theories and the South," *Journal of Southern History* 5 (August, 1939): 291–314; Verner W. Crane, *The Southern Frontier, 1670–1732* (Ann Arbor: University of Michigan Press, 1929).

41 Frederick Jackson Turner, *The Frontier in American History* (New York: Henry Holt and Company, 1921), 11.

42 Allan Greer, "Commons and Enclosure in the Colonization of North America," *American Historical Review* 117 (April, 2012): 365–386.

43 Derek Wall, *The Commons in History: Culture, Conflict, and Ecology* (Cambridge: MIT Press, 2014); Richard A. Bartlett, "Frontier Heritage," in *New Encyclopedia of Southern Culture*, vol. 3: *History*, edited by Charles Reagan Wilson (Chapel Hill: University of North Carolina Press, 2014), 104; Pekka Hämäläinen and Samuel Truett, "On Borderlands," *Journal of American History* 98 (September, 2011): 338–361; Robbie Ethridge (ed.), *Mapping the Mississippian Shatter Zone: The Colonial Indian Slave Trade and Regional Instability in the American South* (Lincoln: University of Nebraska Press, 2009).

44 Stewart, "If John Muir had been an Agrarian," 151. (Italics in the original.)

45 John Majewski and Viken Tchakerian, "The Environmental Origins of Shifting Cultivation: Climates, Soils, and Disease in the Nineteenth-Century US South," *Agricultural History* 81 (fall, 2007): 540.

46 Drew Addison Swanson, "Fighting over Fencing: Agricultural Reform and Antebellum Efforts to Close the Virginia Open Range," *Virginia Magazine of History and Biography* 117 (2010): 104–139; Grady McWhiney, *Cracker Culture: Celtic Ways in the Old South* (Tuscaloosa: University of Alabama Press, 1989); Sam Hilliard, *Hog Meat and Hoecake: Food Supply in the Old South, 1840–1860* (Carbondale: Southern Illinois University Press, 1972); Frank Owsley, *Plain Folk of the Old South* (Baton Rouge: Louisiana State University Press, 1982). To learn more about the environmental history of slavery in the region, see: Thomas G. Andrews, "Beasts of the Southern Wild: Slaveholders, Slaves, and Other Animals in Charles Ball's Slavery in the United States," in *Rendering Nature: Animals, Bodies, Places, and Politics*, edited by Marguerite S. Shaffer and Phoebe S. K. Young (Philadelphia: University of Pennsylvania Press, 2015), 21–47; Mart A. Stewart, "From King Cane to King Cotton: Razing Cane in the Old South," *Environmental History* 12 (January, 2007): 59–79; Mart A. Stewart, "Slavery and the Origins of African American Environmentalism," in *To Love the Wind and the Rain: African Americans and Environmental History*, edited by Dianne D. Glave and Mark Stoll (Pittsburgh: University of Pittsburgh Press, 2005), 9–20; Judith A. Carney, *Black Rice: The African Origins of Rice Cultivation in the Americas* (Cambridge: Harvard University Press, 2002); Stewart, "'Let Us Begin with the Weather,'" 240–256.

47 R. Douglas Hurt, *Agriculture and the Confederacy: Policy, Productivity, and Power in the Civil War South* (Chapel Hill: University of North Carolina Press, 2015); Brian Allen Drake (ed.), *The Blue, the Gray, and the Green: Toward an Environmental History of the Civil War* (Athens: University of Georgia Press, 2015); Kathryn Shively Meier, *Nature's Civil War: Common Soldiers and the Environment in 1862 Virginia* (Chapel Hill: University of North Carolina Press, 2013); Lisa Brady, *War upon the Land: Military Strategy and the Transformation of Southern Landscapes during the American Civil War* (Athens: University of Georgia Press, 2012); Megan Kate

Nelson, *Ruin Nation: Destruction and the American Civil War* (Athens: University of Georgia Press, 2012); Lisa Brady, "From Battlefield to Fertile Ground: The Development of Civil War Environmental History," *Civil War History* 58 (September, 2012): 305–321; Kelby Ouchley, *Flora and Fauna of the Civil War: An Environmental Reference Guide* (Baton Rouge: Louisiana State University Press, 2010); Ann Norton Greene, "War Horses: Equine Technology in the Civil War," in *Industrializing Organisms: Introducing Evolutionary History*, edited by Susan R. Schrepfer and Philip Scranton (New York: Routledge: 2004), 143–166; Mark Fiege, "Gettysburg and the Organic Nature of the American Civil War," in *Natural Enemy, Natural Ally: Toward an Environmental History of Warfare*, edited by Richard P. Tucker and Edmund Russell (Corvallis: Oregon State University, 2004), 93–109.

48 R. Ben Brown, "Free Men and Free Pigs: Closing the Southern Range and the American Property Tradition," *Radical History Review* 108 (2010): 117–137; Shawn Everett Kantor, *Politics and Property Rights: The Closing of the Open Range in the Postbellum South* (Chicago: University of Chicago Press, 1998); Sarah Velody, "The Stock Law Controversy in South Carolina, 1874–1882," Master's Thesis, University of South Carolina, 1990; Steven Hahn, *Roots of Southern Populism: Yeoman Farmer and the Transformation of the Georgia Upcountry, 1850–1890* (Oxford: Oxford University Press, 1983); J. Crawford King, Jr., "The Closing of the Southern Range: An Exploratory Study," *Journal of Southern History* 48 (February, 1982): 53–70.

49 Pete Daniels, *Standing at the Crossroads: Southern Life in the Twentieth Century* (Baltimore: Johns Hopkins University Press, 1996); Edward Ayers, *The Promise of the New South: Life after Reconstruction* (New York: Oxford University Press, 1992); Jack Temple Kirby, *Rural Worlds Lost: The American South, 1920–1960* (Baton Rouge: Louisiana State University Press, 1986). On the rise of industrial farming in the Southeast, see: Monica Gisolfi, "From Crop Lien to Contract Farming: the Roots of Agribusiness in the American South," *Agricultural History* 80 (spring, 2006): 167–189; Michael D. Thompson, "High on the Hog: Swine as Culture and Commodity in Eastern North Carolina," PhD Dissertation, Miami University, 2000.

50 For more on the problematic use of labels in American history, see: James H. Merrell, "Second Thoughts on Colonial Historians and American Indians," *William and Mary Quarterly* 69 (July, 2012): 451–512; James H. Merrell, "Coming to Terms with Early America," *William and Mary Quarterly* 69 (July, 2012): 535–540; Mark Peterson, "Indians and the National Narrative: The Trouble with Words and with Us," *William and Mary Quarterly* 69 (July, 2012): 531–534.

## 2 Making and Breaking Acquaintances

1 Mark A. Maslin, "East African Climate Pulses and Early Human Evolution," *Quaternary Science Reviews* 101 (2014): 1–17; Osbjorn M. Pearson, "Africa: The Cradle of Modern People," in *The Origins of Modern Humans: Biology Reconsidered*, edited by Fred H. Smith and James C. M. Ahern (Hoboken:

Wiley, 2013): 1–43; Sohini Ramachandran, et al., "Support from the Relationship of Genetic and Geographic Distance in Human Populations for a Serial Founder Effect Originating in Africa," *Proceedings of the National Academy of Sciences* 102 (November 1, 2005): 15942–15947; Ian McDougall, et al., "Stratigraphic Placement and Age of Modern Humans from Kibish, Ethiopia," *Nature* 433 (February 17, 2005): 733–736; Tim D. White, et al., "Pleistocene *Homo sapiens* from Middle Awash, Ethiopia," *Nature* 423 (June 12, 2003): 742–747; Chris Stringer, "Human Evolution: Out of Ethiopia," *Nature* 423 (June 12, 2003): 692–695; Ann Gibbons, "Oldest Members of *Homo sapiens* Discovered in Africa," *Science* 300 (June 13, 2003): 1641. Most scientists accept that *Homo sapiens* first evolved in eastern Africa, but some hold dissenting opinions. For examples, see: Brenna M. Henn, et al., "Hunter-Gatherer Genomic Diversity Suggests a Southern African Origin for Modern Humans," *Proceedings of the National Academy of Sciences* 108 (2011): 5154–5162; Y. M. Hou and L. X. Zhao, "An Archaeological View for the Presence of Early Humans in China," *Quaternary International* 223–224 (2010): 10–19; Vinayak Eswaran, et al., "Genomics Refutes an Exclusively African Origin of Humans," *Journal of Human Evolution* 49 (2005): 1–18.

2 Christian A. Tryon, et al., "Late Pleistocene Age and Archaeological Context for the Hominin Calvaria from GvJm-22 (Lukenya Hill, Kenya)," *Proceedings of the National Academy of Sciences* 112 (March 3, 2015): 2682–2687; Eleanor M. L. Scerri, et al., "Earliest Evidence for the Structure of *Homo sapiens* Populations in Africa," *Quaternary Science Reviews* 101 (2014): 207–216; Teresa Rito, et al., "The First Modern Human Dispersals across Africa," *PLOS ONE* 8 (November, 2013): 1–16; Erika Check Hayden, "African Genes Tracked Back," *Science* 500 (August 29, 2013): 514; Anders Eriksson, et al., "Late Pleistocene Climate Change and the Global Expansion of Anatomically Modern Humans," *Proceedings of the National Academies of Science* 109 (October 2, 2012): 16089–16094; Sally C. Reynolds and Andrew Gallagher (eds.), *African Genesis: Perspectives on Hominin Evolution* (Cambridge: Cambridge University Press, 2012); Michael F. Hammer, et al., "Genetic Evidence for Archaic Admixture in Africa," *Proceedings of the National Academy of Sciences* 108 (September 13, 2011): 15123–15128.

3 Daniel Shriner, et al., "Genome-wide Genotype and Sequence-based Reconstruction of the 140,000-year History of Modern Human Ancestry," *Scientific Reports* 4 (2014): 1–9.

4 Ash Parton, et al., "Alluvial Fan Records from Southeast Arabia Reveal Multiple Windows for Human Dispersal," *Geology* 43 (2015): 295–299; Simon J. Armitage, et al., "The Southern Route 'Out of Africa': Evidence for an Early Expansion of Modern Humans into Arabia," *Science* 331 (January 28, 2011): 453–456; Andrew Lawler, "Did Modern Humans Travel Out of Africa via Arabia?" *Science* 331 (January 28, 2011): 387.

5 Some researchers think that this early pulse may have reached China around 100,000 years ago. For more on this incredibly dynamic (and frustratingly opaque) period in human history, see: Wu Liu, et al., "The Earliest Unequivocally Modern Humans in Southern China," *Nature* 526 (October 29, 2015): 696–699; Nicole Boivin, et al., "Human Dispersal Across

Diverse Environments of Asia During the Upper Pleistocene," *Quaternary International* 300 (2013): 32, 47; Andrew Lawler, "Green Arabia," *Science* 345 (August 29, 2014): 994; Eriksson, et al., "Late Pleistocene Climate Change and the Global Expansion of Anatomically Modern Humans," 16090; Hua Liu, et al., "A Geographically Explicit Genetic Model of Worldwide Human-Settlement History," *American Journal of Human Genetics* 79 (August 2006): 235.

6 Marjolein D. Bosch, et al., "New Chronology for Ksar Akil (Lebanon) Supports Levantine Route of Modern Human Dispersal into Europe," *Proceedings of the National Academy of Sciences* 112 (June 23, 2015): 7683–7688; Luca Pagani, et al., "Tracing the Route of Modern Humans Out of Africa by Using 225 Human Genome Sequences from Ethiopians and Egyptians," *American Journal of Human Genetics* 96 (June 4, 2015): 986–991; Israel Hershkovitz, et al., "Levantine Cranium from Manot Cave (Israel) Foreshadows the First European Modern Humans," *Nature* 520 (April 9, 2015): 216–219; Tim Appenzeller, "Human Migrations: Eastern Odyssey," *Nature* 485 (May 2, 2012): 24–26; Morten Rasmussen, et al., "An Aboriginal Australian Genome Reveals Separate Human Dispersals into Asia," *Science* 334 (October 11, 2011): 94–98; P. A. Underhill, et al., "The Phylogeography of Y Chromosome Binary Haplotypes and the Origins of Modern Human Populations," *Annals of Human Genetics* 65 (2001): 43–62.

7 Miikka Tallavaara, et al., "Human Population Dynamics in Europe over the Last Glacial Maximum," *Proceedings of the National Academy of Sciences* 112 (July 7, 2015): 8232–8237; Nicholas J. Conard and Michael Bolus, "Chronicling Modern Humans' Arrival in Europe," *Science* 348 (May 15, 2015): 754–756; S. Benazzi, et al., "The Makers of the Protoaurignacian and Implications for Neanderthal Extinction," *Science* 348 (May 15, 2015): 793–796; Philip R. Nigst, et al., "Early Modern Human Settlement of Europe North of the Alps Occurred 43,500 Years Ago in a Cold Steppe-Type Environment," *Proceedings of the National Academy of Sciences* 111 (October 7, 2014): 14394–14399; Quiaomei Fu, "Genome Sequence of a 45,000-Year-Old Modern Human from Western Siberia," *Nature* 514 (2014): 445–449; J. R. Stewart and C. B. Stringer, "Human Evolution Out of Africa: the Role of Refugia and Climate Change," *Science* 335 (2012): 1317–1321; Tom Higham, et al., "The Earliest Evidence for Anatomically Modern Humans in Northwestern Europe," *Nature* 479 (November 24, 2011): 521–524. Some believe that humans reached Australia as early as 62,000 years ago, but that assertion is contested. See: Morten Rasmussen, et al., "An Aboriginal Australian Genome Reveals Separate Human Dispersals into Asia," *Science* 334 (2011): 94–98; Paul Mellars, et al., "Genetic and Archaeological Perspectives on the Initial Modern Human Colonization of Southern Asia," *Proceedings of the National Academy of Sciences* 110 (June 25, 2013): 10699–10704; M. V. Anikovich, et al., "Early Upper Paleolithic in Eastern Europe and Implications for the Dispersal of Modern Humans," *Science* 315 (2007): 223–226; Yaroslav V. Kuzmin, et al., "The Oldest Directly-Dated Human Remains in Siberia: AMS $^{14}$C Age of Talus Bone from the Baigara Locality, West Siberian Plain," *Journal of Human Evolution* 57 (2009): 91–95.

8 Justin C. Tackney, et al., "Two Contemporaneous Mitogenomes from Terminal Pleistocene Burials in Eastern Beringia," *Proceedings of the National Academy of Sciences* 112 (November 10, 2015): 13833–13838; Bonnie L. Pitblado, "A Tale of Two Migrations: Reconciling Recent Biological and Archaeological Evidence for the Pleistocene Peopling of the Americas," *Journal of Archaeological Research* 19 (2011): 330–331; Marcus J. Hamilton and Briggs Buchanan, "Archaeological Support for the Three-Stage Expansion of Modern Humans across Northeastern Eurasia and into the Americas," *PLOS ONE* 5 (August, 2010): 1–9; Ted Goebel, et al., "The Late Pleistocene Dispersal of Modern Humans in the Americas," *Science* 319 (March 14, 2008): 1497–1502; Erika Tamm, et al., "Beringian Standstill and Spread of Native American Founders," *PLOS ONE* 9 (September, 2007): 1–6.

9 Joaquin Rodriguez-Vidal, "A Rock Engraving made by Neanderthals in Gibraltar" *Proceedings of the National Academy of Sciences* 111 (September 16, 2014): 13301–13306; Karen Hardy, et al., "Neanderthal Medics: Evidence for Food, Cooking, and Medicinal Plants Entrapped in Dental Calculus," *Naturwissenschaften* 99 (August, 2012): 617–626; J. J. Shea, "The Origins of Lithic Projectile Point Technology: Evidence from Africa, the Levant, and Europe," *Journal of Archaeological Science* 33 (June, 2006): 823–846.

10 Susanna Sawyer, et al., "Nuclear and Mitochondrial DNA Sequences from Two Denisovan Individuals," *Proceedings of the National Academy of Sciences* 112 (PNAS Early Edition, 2015): 1–5; Quiaomei Fu, "An Early Modern Human from Romania with a Recent Neanderthal Ancestor," *Nature* 524 (2015): 1–15; Chris Stringer, *Lone Survivors: How We Came to Be the Only Humans on Earth* (New York: Henry Holt & Co., 2012); Yuval Harari, *Sapiens: A Brief History of Humankind* (New York: Harper, 2015); Tom Higham, et al., "The Timing and Spatiotemporal Patterning of Neanderthal Disappearance," *Nature* 512 (August 21, 2014): 306–309; Konrad Lohse and Laurent A. F. Frantz, "Neanderthal Admixture in Eurasia Confirmed by Maximum-Likelihood Analysis of Three Genomes," *Genetics* 196 (April, 2014): 1241–1251; Sriram Sanararaman, et al., "The Genomic Landscape of Neanderthal Ancestry in Present-Day Humans," *Nature* 507 (March 20, 2014): 354–357; Benjamin Vernot and Joshua M. Akey, "Resurrecting Surviving Neanderthal Lineages from Modern Human Genomes," *Science* 343 (February 28, 2014): 1017–1021; Kay Prüfer, et al., "The Complete Genome Sequence of a Neanderthal from the Altai Mountains," *Nature* 505 (January 2, 2014): 43–49; Ewen Callaway, "Hominin DNA Baffles Experts," *Nature* 504 (December 5, 2013): 16–17; Pontus Skoglund and Mattias Jakobsson, "Archaic Human Ancestry in East Asia," *Proceedings of the National Academy of Sciences* 108 (November 8, 2011): 18301–18306; Ann Gibbons, "Who Were the Denisovans?" *Science* 333 (August 26, 2011): 1084–1087; David Reich, "Genetic History of an Archaic Hominin Group from Denisova Cave in Siberia," *Nature* 468 (December 23/30, 2010): 1053–1060; Clive Finlayson, et al., "Late Survival of Neanderthals at the Southernmost Extreme of Europe," *Nature* 443 (October 19, 2006): 850–853. One recent book hypothesizes that dogs helped humans drive Neanderthals extinct. Pat Shipman, *The Invaders: How Humans and Their Dogs Drove Neanderthals to Extinction* (Cambridge: Harvard University Press, 2015).

11 Alon Keinan, et al., "Measurement of the Human Allele Frequency Spectrum Demonstrates Greater Genetic Drift in East Asians than in Europeans," *Nature Genetics* 39 (October, 2007): 1251–1255.
12 Morten Rasmussen, et al., "An Aboriginal Australian Genome Reveals Separate Human Dispersals into Asia," *Science* 334 (2011): 94–98.
13 Ryan N. Gutenkunst, et al., "Inferring the Joint Demographic History of Multiple Populations from Multidimensional SNP Frequency Data," *PLOS Genetics* 5 (October, 2009): 1–11.
14 Daniel Garrigan, et al., "Inferring Human Population Sizes, Divergence Times and Rates of Gene Flow from Mitochondrial, X and Y Chromosome Resequencing Data," *Genetics* 177 (December, 2007): 2195–2207.
15 Ilan Gronau, et al., "Bayesian Inference of Ancient Human Demography from Individual Genome Sequences," *Nature Genetics* 43 (October, 2011): 1031–1034.
16 Andaine Seguin-Orlando, et al., "Genomic Structure in Europeans Dating Back at Least 36,200 Years," *Science* 346 (November 28, 2014): 113–118.
17 Jinchuan Xing, "Genetic Diversity in India and the Inference of Eurasian Population Expansion," *Genome Biology* 11 (2010): 1–13.
18 Qiaomei Fu, et al., "DNA Analysis of an Early Modern Human from Tianyuan Cave, China," *Proceedings of the National Academy of Sciences* 110 (February 5, 2013): 2223–2227.
19 Daniel Shriner, et al., "Genome-Wide Genotype and Sequence-Based Reconstruction of the 140,000 Year History of Modern Human Ancestry," *Scientific Reports* 4 (2014): 1–9.
20 Aylwyn Scally and Richard Durbin, "Revising the Human Mutation Rate: Implications for Understanding Human Evolution," *Nature Reviews Genetics* 13 (October, 2012): 745–753.
21 On the nonexistence of human races, see: Michael Yudell, *Race Unmasked: Biology and Race in the 20th Century* (New York: Columbia University Press, 2014); Robert Wald Sussman, *The Myth of Race: The Troubling Persistence of an Unscientific Idea* (Cambridge: Harvard University Press, 2014); Jeffrey C. Long and Rick A. Kittles, "Human Genetic Diversity and the Nonexistence of Biological Races," *Human Biology* 75 (August, 2003): 449–471; Frank B. Livingstone and Theodosius Dobzhansky, "On the Non-Existence of Human Races," *Current Anthropology* 3 (June, 1962): 279–281. On the position that biological races *are* real, see: Nicholas Wade, *A Troublesome Inheritance: Genes, Race, and Human History* (New York: Penguin, 2014). On the "labile" nature of the human genome, see: Kelley Harris, "Evidence for Recent, Population-Specific Evolution of the Human Mutation Rate," *Proceedings of the National Academy of Sciences* 112 (March 17, 2015): 3439–3444.
22 M. Gallego Llorente, et al., "Ancient Ethiopian Genome Reveals Extensive Eurasian Admixture throughout the African Continent," *Science* 350 (October 8, 2015): 1–7; Deepti Gurdasani, et al., "The African Genome Variation Project Shapes Medical Genetics in Africa," *Nature* 517 (January 15, 2015): 327–332; Jason A. Hodgson, "Early Back-to-Africa Migration into the Horn of Africa," *PLOS Genetics* 10 (June, 2014): 1–18.

23 Garrett Hellenthal, et al., "A Genetic Atlas of Human Admixture History," *Science* 343 (February 14, 2014): 747–751.
24 Ernst Mayr, "Lamarck Revisited," *Journal of the History of Biology* 5 (Spring, 1972): 66–67; Charles Darwin, *The Variation of Animals and Plants under Domestication, Volume I* (London: John Murray, 1868): 26.
25 Edmund Russell, *Gambling on Evolution: the Coevolution of English Culture with Fierce, Fleet, and Fancy Dogs* (New York: Cambridge University Press, forthcoming).
26 Juliette Clutton-Brock, "Origins of the Dog: Domestication and Early History," in *The Domestic Dog: Its Evolution, Behaviour, and Interactions with People*, edited by James Serpell (Cambridge: Cambridge University Press, 1995), 12; Simon J. M. Davis and Francois R. Valla, "Evidence for Domestication of the Dog 12,000 Years Ago in the Natufian of Israel," *Nature* 276 (December 7, 1978): 608–610.
27 Carles Vilà, et al., "Multiple and Ancient Origins of the Domestic Dog," *Science* 276 (June 13, 1997): 1687–1689.
28 Guo-dong Wang, et al., "The Genomics of Selection in Dogs and the Parallel Evolution between Dogs and Humans," *Nature Communications* 4 (May 13, 2013): 1–9.
29 Pontus Skoglund, et al., "Ancient Wolf Genome Reveals an Early Divergence of Domestic Dog Ancestors and Admixture into High-Latitude Breeds," *Current Biology* 25 (2015): 1–5; Anna S. Druzhkova, et al., "Ancient DNA Analysis Affirms the Canid from Altai as a Primitive Dog," *PLOS ONE* 8 (March, 2013): 1–6; Pontus Skoglund, et al., "Estimation of Population Divergence Times from Non-Overlapping Genomic Sequences: Examples from Dogs and Wolves," *Molecular Biology and Evolution* 28 (2011): 1505–1517. It should be noted, however, that many geneticists continue to insist that dogs were domesticated around 16,000 years ago. Adam H. Freedman, et al., "Genome Sequencing Highlights the Dynamic Early History of Dogs," *PLOS Genetics* 10 (January, 2014): 1–12.
30 Mietje Germonpré, et al., "Palaeolithic Dog Skulls at the Gravettian Předmostí site, the Czech Republic," *Journal of Archaeological Science* 39 (2012): 184–202; Nikolai D. Ovodov, et al., "A 33,000-Year-Old Incipient Dog from the Altai Mountains of Siberia: Evidence of the Earliest Domestication Disrupted by the Last Glacial Maximum," *PLoS ONE* 6 (July, 2011): 1–7; Mietje Germonpré, et al., "Fossil Dogs and Wolves from Palaeolithic Sites in Belgium, the Ukraine and Russia: Osteometry, Ancient DNA and Stable Isotopes," *Journal of Archaeological Science* 36 (2009): 473–490. It is worth pointing out that some archaeologists remain skeptical toward such ancient claims. For example, see: Hannes Napierala and Hans-Peter Uepmann, "A 'New' Palaeolithic Dog from Central Europe," *International Journal of Osteoarchaeology* 22 (2012): 127–137.
31 Laura M. Shannon, et al., "Genetic Structure in Village Dogs Reveals a Central Asian Domestication Origin," *Proceedings of the National Academy of Sciences* 112 (November 3, 2015): 13639–13644; David Grimm, "Dawn of the Dog," *Science* 348 (April 17, 2015): 274–279; O. Thalmann, et al., "Complete Mitochondrial Genomes of Ancient Canids Suggest a European

Origin of Domestic Dogs," *Science* 342 (2013): 871–874; A. K. Niskanen, "MHC Variability Supports Dog Domestication from a Large Number of Wolves: High Diversity in Asia," *Heredity* 110 (2013): 80–85; Z. L. Ding, et al., "Origins of Domestic Dog in Southern East Asia is Supported by Analysis of Y-Chromosome DNA" *Heredity* 108 (2012): 507–514; Jun-Feng Pang, et al., "mtDNA Data Indicate a Single Origin for Dogs South of Yangtze River, Less Than 16,300 Years Ago, from Numerous Wolves" *Molecular Biology and Evolution* 26 (2009): 2849–2864; Bridgett M. vonHoldt, "Genome-Wide SNP and Haplotype Reveal a Rich History Underlying Dog Domestication," *Nature* 464 (April 8, 2010): 898; Peter Savolainen, et al., "Genetic Evidence for an East Asian Origin of Domestic Dogs," *Science* 298 (November 22, 2002): 1610–1613.

32 Diane Gifford-Gonzalez and Olivier Hanotte, "Domesticating Animals in Africa: Implications of Genetic and Archaeological Findings," *Journal of World Prehistory* 24 (2011): 15; Adam R. Boyko, et al., "Complex Population Structure in African Village Dogs and Its Implications for Inferring Dog Domestication History," *Proceedings of the National Academy of Sciences* 106 (August 18, 2009): 13903–13908.

33 Germonpré, et al., "Fossil dogs and wolves from Palaeolithic sites in Belgium, the Ukraine and Russia," 473–490; Germonpré et al., "Palaeolithic dog skulls at the Gravettian Předmostí site, the Czech Republic," 184–202.

34 Juliet Clutton-Brock, *Natural History of Domesticated Mammals*, second edition (Cambridge: Cambridge University Press, 1999), 4; Rudolf Musil, "Evidence for the Domestication of Wolves in Central European Magdalenian Sites," in *Dogs through Time: an Archaeological Perspective Proceedings of the 1st ICAZ Symposium on the History of the Domestic Dog*, edited by Susan J. Crockford (Oxford: British Archaeological Reports International Series 889, 2000), 21.

35 M. F. Ashley Montagu, "On the Origin of the Domestication of the Dog," *Science* 96 (July 31, 1942): 111–112.

36 Raymond Coppinger and Lorna Coppinger, *Dogs: A Startling New Understanding of Canine Origin, Behavior, and Evolution* (New York: Scribner, 2001), 60; Edmund Russell, *Evolutionary History: Uniting History and Biology to Understand Life on Earth* (New York: Cambridge University Press, 2011), 60.

37 Melinda A. Zeder, "The Domestication of Animals," *Journal of Anthropological Research* 68 (summer, 2012): 172.

38 Pierre Ducos, "Defining Domestication: a Clarification," in *The Walking Larder: Patterns of Domestication, Pastoralism, and Predation*, edited by Juliet Clutton-Brock (New York: Routledge, 1990), 29. See also: Sandor Bokonyi, "Definitions of Animal Domestication," in *The Walking Larder: Patterns of Domestication, Pastoralism, and Predation*, edited by Juliet Clutton-Brock (New York: Routledge, 1990), 24.

39 Edmund Russell, "Coevolutionary History," *American Historical Review* 119 (December, 2014): 1514–1528; Russell, *Evolutionary History*, 69: 69. See also: Michael Pollan, *Botany of Desire: A Plant's-Eye View of the World* (New York: Random House, 2001), xxi; T. P. O'Connor, "Working at Relationships: Another Look at Animal Domestication," *Antiquity* 71 (March, 1997):

149–156. See also: Nerissa Russell, *Social Zooarchaeology: Humans and Animals in Prehistory* (Cambridge: Cambridge University Press, 2012): 215; Coppinger and Coppinger, *Dogs*, 41, 56.
40 Monique A. R. Udell, et al., "What Did Domestication Do to Dogs? A New Account of Dogs' Sensitivity to Human Actions," *Biological Reviews* 85 (2010): 328. See also: Miho Nagasawa, et al., "Oxytocin-gaze Positive Loop and the Coevolution of Human-Dog Bonds," *Science* 348 (April 17, 2015): 333–336; Ádám Miklósi and József Topál, "What Does it Take to Become 'Best Friends'? Evolutionary Changes in Canine Social Competence," *Trends in Cognitive Sciences* 17 (June, 2013): 287–294; Brian Hare, et al., "The Domestication of Social Cognition in Dogs," *Science* 298 (November 22, 2002): 1634–1636;
41 Domestic animals are generally much smaller than their wild (never-domesticated) conspecifics. Other telltale signs of domestication include shorter snouts and crowded teeth.
42 Anna V. Kukekova, et al., "Mapping Loci for Fox Domestication: Deconstruction/Reconstruction of a Behavioral Phenotype," *Behavior Genetics* 41 (July, 2011): 593–606; L. Trut, et al., "Animal Evolution During Domestication: The Domesticated Fox as a Model," *Bioessays* 31 (2009): 349–360; I. E. Elia, "A Foxy View of Human Beauty: Implications of the Farm Fox Experiment for Understanding the Origins of Structural and Experiential Aspects of Facial Attractiveness," *Quarterly Review of Biology* 88 (September, 2013): 163–183 For the best popular account of this novel experiment, see: Evan Ratliff, "Taming the Wild," *National Geographic* 220 (March, 2011): 34–59.
43 Zeder, "The Domestication of Animals," 169.
44 Adam S. Wilkins, "Domestication Syndrome," *Genetics* 197 (July, 2014): 795–808. Some have even claimed that these morphological changes also occurred in humans, and that we are no less domesticated than other animals. Ann Gibbons, "How We Tamed Ourselves – and Became Modern," *Science* 346 (October, 24, 2014): 405–406; Helen M. Leach, "Human Domestication Reconsidered," *Current Anthropology* 44 (June, 2003): 349–368.
45 Roderick F. Nash, "Island Civilization: A Vision for Human Occupancy of Earth in the Fourth Millennium," *Environmental History* 15 (July, 2010): 373.
46 Juliet Clutton-Brock, *Natural History of Domesticated Mammals*, second edition (Cambridge: Cambridge University Press, 1999), 30.
47 Fiona B. Marshall, et al., "Evaluating the roles of directed breeding and gene flow in animal domestication," *Proceedings of the National Academy of Sciences* 111 (April 29, 2014): 6153–6158; Vilà, et al., "Multiple and Ancient Origins of the Domestic Dog," 1689; Germonpré, et al., "Palaeolithic Dog Skulls at the Gravettian Předmostí Site," 195–196. Some believe that we should reserve use of the label "domestication" to the initial union between species. Greger Larson, et al., "Current Perspectives and the Future of Domestication Studies," *Proceedings of the National Academy of Sciences* 111 (April 29, 2014): 6140.
48 Ovodov, et al., "A 33,000-Year-Old Incipient Dog from the Altai Mountains of Siberia," 6.
49 Marshall, et al., "Evaluating the Roles of Directed Breeding and Gene Flow in Animal Domestication," 6153–6158.

50 Bruce D. Smith, "Niche Construction and the Behavioral Context of Plant and Animal Domestication," *Evolutionary Anthropology* 16 (2007): 188–199; Bruce D. Smith, "The Ultimate Ecosystem Engineers," *Science* 315 (March 30, 2007): 1797–1798; Melinda A. Zeder, "The Origins of Agriculture in the Near East," *Current Anthropology* 52 (October, 2011): s221–s235; Jean-Denis Vigne, "The Origins of Animal Domestication and Husbandry: A Major Change in the History of Humanity and the Biosphere," *Comptes Rendus Biologies* 334 (2011): 171–181.

51 Peter Rowley-Conwy, et al., "Distinguishing Wild Boar from Domestic Pigs in Prehistory: A Review of Approaches and Recent Results," *Journal of World Prehistory* 25 (2012): 1–44.

52 Mirte Bosse, et al., "Untangling the Hybrid Nature of Modern Pig Genomes: A Mosaic Derived From Biogeographically Distinct and Highly Divergent *Sus scrofa* Populations" *Molecular Ecology* 23 (2014): 4089–4102; Umberto Albarella, et al., "The Domestication of the Pig (*Sus scrofa*): New Challenges and Approaches," in *Documenting Domestication: New Genetic and Archaeological Paradigms*, edited by Melinda A. Zeder, et al. (Berkeley: University of California Press, 2006): 219.

53 Benjamin Arbuckle, "Data Sharing Reveals Complexity in the Westward Spread of Domestic Animals across Neolithic Turkey," *PLoS One* 9 (June, 2014): 1–11; Melinda A. Zeder, "Pathways to Animal Domestication," in *Biodiversity in Agriculture: Domestication, Evolution, and Sustainability*, edited by P. Gepts, Thomas R. Famula, Robert L. Bettinger, Stephen B. Brush, Ardeshir B. Damania, Patrick E. McGuire, and Calvin O. Qualset (Cambridge: Cambridge University Press, 2012): 241–242.

54 Perista Paschou, et al., "Maritime Route of Colonization of Europe," *Proceedings of the National Academy of Sciences* 111 (June 24, 2014): 9211–9216; Anne Tresset and Jean-Denis Vigne, "Last Hunter-Gatherers and First Farmers of Europe," *Comptes Rendus Biologies* 334 (2011): 184.

55 Tresset and Vigne, "Last Hunter-Gatherers and First Farmers of Europe," 183–184.

56 Ruth Bollongino, "2,000 Years of Parallel Societies in Stone Age Central Europe," *Science* 342 (October 25, 2013): 479–481; Pontus Skoglund, et al., "Origins and Genetic Legacy of Neolithic Farmers and Hunter-Gatherers in Europe," *Science* 336 (April 27, 2012): 469.

57 Greger Larson, et al., "Ancient DNA, Pig Domestication, and the Spread of the Neolithic into Europe," *Proceedings of the National Academy of Sciences* 104 (September 25, 2007): 15276–15281.

58 Meirav Meiri, et al., "Ancient DNA and Population Turnover in Southern Levantine Pigs – Signature of the Sea Peoples Migration?" *Scientific Reports* 3 (2013): 1–8; Claudio Ottoni, et al., "Pig Domestication and Human-Mediated Dispersal in Western Eurasia Revealed through Ancient DNA and Geometric Morphometrics," *Molecular Biology and Evolution* 30 (2012): 824–832; Sam White, "From Capitalist Pigs to Globalized Pig Breeds: A Study in Animal Cultures and Evolutionary History," *Environmental History* 16 (January, 2011): 94–120.

59 Laurent A. Frantz, "Evidence of Long-term Gene Flow and Selection during Domestication from Analyses of Eurasian Wild and Domestic Pig Genomes,"

*Nature Genetics* 47 (August 31, 2015): 1141–1148; Marshall, et al., "Evaluating the roles of directed breeding and gene flow in animal domestication," 6153–6158.
60  Greger Larson, et al., "Worldwide Phylogeography of Wild Boar Reveals Multiple Centers of Pig Domestication," *Science* 307 (March 11, 2005): 1618–1621. Regarding the claim that pigs were also domesticated in Africa, see: Marcel Amills, et al., "Domestic Pigs in Africa," *African Archaeological Review* 30 (2013): 73–82; Juliet Clutton-Brock, *Animals as Domesticates: A World View through History* (East Lansing: Michigan State University Press, 2012), 50; Diane Gifford-Gonzalez and Olivier Hanotte, "Domesticating Animals in Africa: Implications of Genetic and Archaeological Findings," *Journal of World Prehistory* 24 (2011): 13.
61  Martien A. M. Goenen, et al., "Analyses of Pig Genomes Provide Insight into Porcine Demography and Evolution," *Nature* 491 (November 15, 2012): 393–398; Albarella, et al., "The Domestication of the Pig (*Sus scrofa*): New Challenges and Approaches"; Umberto Albarella, et al. (eds.), *Pigs and humans: 10,000 years of Interaction* (New York: Oxford University Press, 2007); Larson, et al., "Worldwide Phylogeography of Wild Boar Reveals Multiple Centers of Pig Domestication," 1618–1621.
62  Marshall, et al., "Evaluating the roles of directed breeding and gene flow in animal domestication," 6153–6158; White, "From Globalized Pig Breeds to Capitalist Pigs," 96; Greger Larson, et al., "Patterns of East Asian pig domestication, migration, and turnover revealed by modern and ancient DNA," *Proceedings of the National Academy of Sciences* 107 (April 27, 2010): 7686–7691; Erin S. Luetkemeir, "Multiple Asian Pig Origins Revealed Through Genomic Analyses," *Molecular Phylogenetics and Evolution* 54 (2010): 680–686; H. J. Megens, et al., "Biodiversity of Pig Breeds from China and Europe Estimated from Pooled DNA Samples," *Genetics Selection Evolution* 40 (2008): 103–128.
63  Tina Hesman Saey, "Ancient Horse's DNA Fills in Picture of Equine Evolution," *Science News* 184 (July 27, 2013): 5–6; Ludovic Orlando, et al., "Recalibrating Equus Evolution Using the Genome Sequence of an Early Middle Pleistocene Horse," *Nature* 499 (July 4, 2013): 74–78; Carles Vila, et al., "Genetic Documentation of Horse and Donkey Domestication," in *Documenting Domestication: New Genetic and Archaeological Paradigms*, edited by Melinda A. Zeder, Daniel G. Bradley, Eve Emshwiller, and Bruce D. Smith (Berkeley: University of California Press, 2006), 344.
64  Alessandro Achilli, et al., "Mitochondrial Genomes from Modern Horses Reveal the Major Haplogroups that Underwent Domestication," *Proceedings of the National Academy of Sciences* 109 (February 14, 2012): 2452.
65  Robin Bendrey, "From Wild Horses to Domestic Horses: A European Perspective," *World Archaeology* 44 (2012): 149–150.
66  Marsha A. Levine, "Domestication and Early History of the Horse," in *The Domestic Horse: The Origins, Development and Management of its Behaviour*, edited by D. S. Mills and S. M. McDonnell (New York: Cambridge University Press, 2005), 7–15.
67  Vila, et al., "Genetic Documentation of Horse and Donkey Domestication," 342.
68  Alan K. Outram, et al., "The Earliest Horse Harnessing and Milking," *Science* 323 (March 6, 2009): 1332–1335. See also: Alan K. Outram, "Patterns of Pastoralism

in Later Bronze Age Kazakhstan: New Evidence from Faunal and Lipid Residue Analyses," *Journal of Archaeological Science* 39 (2012): 2424–2435; Vera Warmuth, et al., "Reconstructing the Origin and Spread of Horse Domestication in the Eurasian Steppe," *Proceedings of the National Academy of Sciences* 109 (May 22, 2012): 8202–8206; Bendrey, "From Wild Horses to Domestic Horses," 136.
69 Sandra L. Olsen, "Early Horse Domestication on the Eurasian Steppe," in *Documenting Domestication: New Genetic and Archaeological Paradigms*, edited by Melinda A. Zeder, Daniel G. Bradley, Eve Emshwiller, and Bruce D. Smith (Berkeley: University of California Press, 2006), 245–269.
70 Arne Ludwig, et al., "Coat Color Variation at the Beginning of Horse Domestication," *Science* 324 (2009): 485.
71 Wolfgang Haak, "Massive Migration from the Steppe was a Source for Indo-European Languages in Europe," *Nature* 522 (June 11, 2015): 207–211; Anne Tresset and Jean-Denis Vigne, "Last Hunter-Gatherers and First Farmers of Europe," *Comptes Rendus Biologies* 334 (2011): 184; David W. Anthony, *The Horse, the Wheel, and Language: How Bronze-Age Riders from the Eurasian Steppes Shaped the Modern World* (Princeton: Princeton University Press, 2010).
72 Vera Warmuth, et al., "Ancient Trade Routes Shaped the Genetic Structure of Horses in Eastern Eurasia" *Molecular Ecology* 22 (2013): 5340–5351; Yuan Jing, "The Origins and Development of Animal Domestication in China," *Chinese Archaeology* 8 (2008): 4; Stanley J. Olsen, "The Horse in Ancient China and its Cultural Influence in Some Other Areas," *Proceedings of the Academy of Natural Sciences of Philadelphia* 140 (1988): 151–189.
73 Olsen, "Early Horse Domestication on the Eurasian Steppe," 251.
74 Leon B. Blair, "The Origin and Development of the Arabian Horse," *Southwestern Historical Quarterly* 68 (January, 1965): 303–316.
75 Kelekna, *The Horse in Human History*, 218; Clutton-Brock, *Animals as Domesticates*, 4.
76 James L. A. Webb, Jr., "The Horse and Slave Trade between the Western Sahara and Senegambia," *Journal of African History* 34 (1993): 221–246; Robin Law, *The Horse in West African History: The Role of the Horse in the Societies of Pre-Colonial Africa* (London: International African Institute, 1980); Robin Law, "Horses, Firearms, and Political Power in Pre-Colonial West Africa," *Past & Present* 72 (August, 1976): 112–132. By comparison, horses did not filter into South Africa until considerably later. See: Sandra Swart, *Riding High: Horses, Humans and History in South Africa* (Johannesburg: Wits University Press, 2010); Greg Bankoff and Sandra Swart, *Breeds of Empire: The 'Invention' of the Horse in Southeast Asia and Southern Africa, 1500–1950* (Copenhagen: Nordic Institute of Asian Studies Press, 2007).
77 Bendrey, "From Wild Horses to Domestic Horses," 149; Achilli, et al., "Mitochondrial Genomes from Modern Horses Reveal the Major Haplogroups that Underwent Domestication," 2452; Vera Warmuth, et al., "European Domestic Horses Originated in Two Holocene Refugia," *PLoS One* 6 (2011): 1–7; Jaime Lira, et al., "Ancient DNA Reveals Traces of Iberian Neolithic and Bronze Age Lineages in Modern Iberian Horses," *Molecular Ecology* 19 (2010): 64–78.

78 Robin Bendrey, "Population Genetics, Biogeography, and Domestic Horse Origins and Diffusions," *Journal of Biogeography* 41 (2014): 1441–1442; Marshall, "Evaluating the Roles of Directed Breeding and Gene Flow in Animal Domestication," 6154; Warmuth, et al., "Reconstructing the Origin and Spread of Horse Domestication in the Eurasian Steppe," 8202; Marsha Levine, "mtDNA and Horse Domestication: The Archaeologist's Cut," in *Equids in Time and Space*, edited by Marjan Mashkour (London: Oxbow Books: 2006), 192; Sandra L. Olsen, "Early Horse Domestication on the Eurasian Steppe," in *Documenting Domestication: New Genetic and Archaeological Paradigms*, edited by Melinda Zeder, Daniel G. Bradley, Eve Emshwiller, and Bruce D. Smith (Berkeley: University of California Press, 2006), 256; Robin Bendrey, "From Wild Horses to Domestic Horses: A European Perspective," *World Archaeology* 44 (2012): 149.

79 Levine, "mtDNA and Horse Domestication," 192–201; Bendrey, "From Wild Horses to Domestic Horses," 150.

80 Jacopo De Grossi Mazzorin and Antonio Tagliacozzo, "Morphological and Osteological Changes in the Dog from the Neolithic to the Roman Period in Italy," in *Dogs through Time: an Archaeological Perspective, Proceedings of the 1st ICAZ Symposium on the History of the Domestic Dog*, edited by Susan J. Crockford (Oxford: British Archaeological Reports International Series 889, 2000), 141; Kate M. Clark, "Dogged Persistence: the Phenomenon of Canine Skeletal Uniformity in British Prehistory," in *Dogs through Time: an Archaeological Perspective, Proceedings of the 1st ICAZ Symposium on the History of the Domestic Dog*, edited by Susan J. Crockford (Oxford: British Archaeological Reports International Series 889, 2000), 163.

81 This was especially true in late-fifteenth-century Iberia, where Christian armies employed dogs in their attack on the Moors. There may have been any number of canine varieties over the years, but the breeds generally fell into one of three distinct categories: *lebrel* (greyhound), *mastín* (mastiff), and *alano* (wolfhound). Charles Hudson, *Knights of Spain, Warriors of the Sun: Hernando de Soto and the South's Ancient Chiefdoms* (Athens: University of Georgia Press, 1997), xvi, 74–75; John G. Varner and Jeannette J. Varner, *Dogs of the Conquest* (Norman: University of Oklahoma Press, 1983), xiv.

82 Regarding village dogs and pariah dogs, see: Shannon, et al., "Genetic Structure in Village Dogs Reveals a Central Asian Domestication Origin," 13639–13644; Boyko, et al., "Complex Population Structure in African Village Dogs and Its Implications for Inferring Dog Domestication History," 13903–13908; I. Lehr Brisbin, "The Pariah: Its Ecology and Importance to the Origin, Development and Study of Pure Bred Dogs," *Pure-Bred Dogs American Kennel Gazette* 94 (1977): 22–29. Regarding the evolutionary history of dingoes, who arrived in Australia between 3,500 and 5,000 years ago, see: Bradley P. Smith and Carla A. Litchfield "A Review of the Relationship between Indigenous Australians, Dingoes (*Canis dingo*) and Domestic Dogs (*Canis familiaris*)," *Anthozoos* 22(2009): 111–128; Savolainen et al., "A Detailed Picture of the Origin of the Australian Dingo, Obtained from the Study of Mitochondrial DNA," *Proceedings of the National Academy of Sciences* 101 (August 17, 2004): 12387–12390.

83 White, "From Globalized Pig Breeds to Capitalist Pigs," 100.

84 White, "From Globalized Pig Breeds to Capitalist Pigs," 96, 99. See also: Rowley-Conwy, et al., "Distinguishing Wild Boar from Domestic Pigs in Prehistory," 10.

## 3 When Ferality Reigned

1 Evidence suggests that people and dogs waited in Beringia approximately 8,000 to 15,000 years. Justin C. Tackney, et al., "Two Contemporaneous Mitogenomes from Terminal Pleistocene Burials in Eastern Beringia," *Proceedings of the National Academy of Sciences* 112 (November 10, 2015): 13833–13838; Maanasa Raghavan, et al., "Genomic Evidence for the Pleistocene and Recent Population History of Native Americans," *Science* 349 (July 21, 2015): 1–20; Erika Tamm, et al., "Beringian Standstill and Spread of Native American Founders," *PLOS ONE* 9 (September, 2007): 1–6.
2 Most researchers agree that dogs accompanied the first human immigrants into North America. Geneticists cite evidence that the dogs who lived in North America prior to European contact were most closely related to dogs in East Asia. What is more, zooarchaeologists have recovered Pleistocene-era dog skeletons from the Kamchatka Peninsula in eastern Siberia, which proves that dogs have an ancient presence in the region. Jennifer A. Leonard, et al., "Ancient DNA Evidence for Old World Origin of New World Dogs," *Science* 298 (November 22, 2002): 1613–1616; Darcy F. Morey, "Burying Key Evidence: the Social Bond Between Dogs and People," *Journal of Archaeological Science* 33 (2006): 166–167. Even so, some researchers think that dogs did not accompany America's first human immigrants, and that they instead arrived alongside later waves. K. E. Witt, et al., "DNA Analysis of Ancient Dogs of the Americas: Identifying Possible Founding Haplotypes and Reconstructing Population Histories," *Journal of Human Evolution* 79 (2015): 105–118.
3 Justin C. Tackney, et al., "Two Contemporaneous Mitogenomes from Terminal Pleistocene Burials in Eastern Beringia," *Proceedings of the National Academy of Sciences* 112 (November 10, 2015): 13833–13838; Pontus Skoglund, "Genetic Evidence for Two Founding Populations of the Americas," *Nature* 525 (September 3, 2015): 104–108; Maanasa Raghavan, et al., "Genomic Evidence for the Pleistocene and Recent Population History of Native Americans," *Science* 349 (August 21, 2015): 841; Anders Eriksson, et al., "Late Pleistocene Climate Change and the Global Expansion of Anatomically Modern Humans," *Proceedings of the National Academies of Science* 109 (October 2, 2012): 16089–16094; Pitblado, "A Tale of Two Migrations," 327–375.
4 Michelle A. Chaput, et al., "Spatiotemporal Distribution of Holocene Populations in North America," *Proceedings of the National Academy of Sciences* 112 (September 29, 2015): 12127–12132.
5 Eriksson, et al., "Late Pleistocene climate change and the global expansion of anatomically modern humans" 16089–16094; Ted Goebel, et al., "The Late Pleistocene Dispersal of Modern Humans in the Americas," *Science* 319 (March 14, 2008): 1501. See also: David G. Anderson, "Paleoindian Occupations in the Southeastern United States," in *New Perspectives on*

*the First Americans*, edited by Bradley T. Lepper and Robson Bonnichsen (College Station: Center for the Study of the First Americans, 2004), 119–128; David G. Anderson and Michael K. Faught, "Paleoindian Artefact Distributions: Evidence and Implications," *Antiquity* 74 (2000): 507–513.

6 Alan Cooper, et al., "Abrupt Warming Events Drove Late Pleistocene Holarctic Megafaunal Turnover," *Science* 349 (August 7, 2015): 602–606; Christopher Sandom, et al., "Global late Quaternary Megafauna Extinctions Linked to Humans, not Climate Change," *Proceedings of the Royal Society B* 281 (July, 2014): 1–9; Graham W. Prescotta, et al., "Quantitative Global Analysis of the Role of Climate and People in Explaining Late Quaternary Megafaunal Extinctions," *Proceedings of the National Academy of Sciences* 109 (March 20, 2012): 4527–4531; J. Tyler Faith and Todd A. Surovell, "Synchronous Extinction of North America's Pleistocene Mammals," *Proceedings of the National Academy of Sciences* 106 (December 8, 2009): 20641–20645; Paul L. Koch and Anthony D. Barnosky, "Late Quaternary Extinctions: State of the Debate," *Annual Review of Ecology, Evolution, and Systematics* 37 (2006): 215–250; R. Dale Guthrie "New Carbon Dates Link Climatic Change with Human Colonization and Pleistocene Extinctions," *Nature* 441 (May 11, 2006): 207–209; David A. Burney and Timothy F. Flannery, "Fifty Millennia of Catastrophic Extinctions After Human Contact" *Trends in Ecology and Evolution* 20 (July, 2005): 395–401; Stuart Fiedel and Gary Haynes, "A Premature Burial: Comments on Grayson and Meltzer's 'Requiem for Overkill,'" *Journal of Archaeological Science* 31 (2004): 121–131; Donald K. Grayson and David J. Meltzer, "A Requiem for North American Overkill," *Journal of Archaeological Science* 30 (2003): 585–593; John E. Guilday, "Appalachia 11,000–12,000 Years Ago: A Biological Review," *Archaeology of Eastern North America* 10 (fall, 1982): 22–26; C. Turner, "Teeth, Needles, Dogs, and Siberia: Bioarchaeological Evidence for the Colonization of the New World," in *The First Americans: The Pleistocene Colonization of the New World*, edited by N. Jablonski (San Francisco: California Academy of Sciences, 2002), 146. Some have even suggested that an extraterrestrial impact may have wiped out the Pleistocene megafauna in North America, but that claim is disputed. R. B. Firestone, et al., "Evidence for an Extraterrestrial Impact 12,900 Years Ago That Contributed to the Megafaunal Extinctions and the Younger Dryas Cooling," *Proceedings of the National Academy of Sciences* 104 (October 9, 2007): 16016–16021; David J. Meltzer, et al., "Chronological Evidence Fails to Support Claim of an Isochronous Widespread Layer of Cosmic Impact Indicators Dated to 12,800 Years Ago," *Proceedings of the National Academy of Sciences* 111 (May 27, 2014): e2162–e2171.

7 David G. Anderson, et al., "Pleistocene Human Settlement in the Southeastern United States: Current Evidence and Future Directions," *PaleoAmerica* 1 (January, 2015): 7–51.

8 V. P. Steponaitis, "Prehistoric Archaeology in the Southeastern United States, 1970–1985," *Annual Review of Anthropology* 15 (October, 1986): 367.

9 Guilday, "Appalachia 11,000–12,000 years ago," 22–23.

10 Donald Davis (ed.), *Southern United States: An Environmental History* (Santa Barbara: ABC-CLIO, 2006), 18.

11 Lynn M. Snyder and Jennifer A. Leonard, "The Diversity and Origin of American Dogs," in *Subsistence Economies of Indigenous North American Societies: A Handbook*, edited by Bruce D. Smith (Washington, DC: Smithsonian Institution Scholarly Press, 2011), 525–542.
12 Most studies that purport to examine prehistoric dogs in southeastern North America actually examine dogs found at sites west of the Appalachian Mountains. Renee B. Walker, et al., "Early and Mid-Holocene Dogs in Southeastern North America: Examples from Dust Cave," *Southeastern Archaeology* 24 (summer, 2005): 83–92.
13 David G. Anderson, "Dogs from the G.S. Lewis West Site (38Ak228)," Ms. on file, Savannah River Archaeological Research Program, South Carolina Institute of Archaeology and Anthropology, Columbia, South Carolina (n. d.); Clarence B. Moore, *The Georgia and South Carolina Coastal Expeditions* (Tuscaloosa: University of Alabama Press, 1998), 41, 44, 71, 73, 74, 213; Jessica Zimmer; "Native Americans' Treatment of Dogs in Prehistoric and Historic Florida," Master's Thesis, Florida State University, Tallahassee, Florida (2007), 52. Regrettably, scientists have yet to recover conclusive evidence of dogs in the South from the Paleoindian Era. David Webb and Erika Simons report that numerous Paleoindian sites along the Aucilla River in northern Florida contain disarticulated bone fragments that *might* have belonged to domestic dogs, and, if true, these chipped bones would represent the earliest evidence of domestic animals in the South. Even so, Webb and Simons hasten caution without further evidence. Evidence of dogs in the Archaic Era (8,000 years ago to 3,000 years ago) is only slightly more promising. The earliest probable trace of dogs in the South hails from Brevard County, Florida. The evidence consists of numerous tools, carved from canid bones and found buried with humans, which have been dated to more than 7,500 years old. The sheer number of these bone tools, including their proximity to humans, renders it likely that they came from dogs and not wolves, but the latter explanation cannot be excluded. Similar canid-bone tools have turned up on the banks of the Savannah River in the South Carolina Lowcountry, and have been dated to the same period. Meanwhile, other Archaic-Era sites offer similarly ambiguous discoveries. Not too far from Sarasota, Florida, archaeologists have unearthed a canid's canine tooth that bears decorative engraving and which evidently served as a pendant of sorts. It is difficult to say with any definitive proof that the tooth belonged to a dog and not a wolf, but the engravings obviously suggest some kind of cultural affiliation. Finally, some sites indicate that southeastern dogs may have also served gastronomic purposes during the Archaic Era. For example, Jerald Milanich cites the disarticulated nature of canid bones found in refuse pits near the St. Johns River as evidence that the region's inhabitants ate dogs during the Archaic Era. David Webb and Erika Simons, "Vertebrate Paleontology," in *First Floridians and Last Mastodons: The Page-Ladson Site in the Aucilla River*, edited by David Webb (Dordrecht, Netherlands: Springer, 2006), 548; Jerald T. Milanich, *Archaeology of Pre-Columbian Florida* (Gainesville: University of Florida Press, 1994); James B. Stoltman, *Groton Plantation: An Archaeological Study of a South Carolina Locality* (Cambridge: Peabody Museum of Archaeology and Ethnology, 1974).

14 Jeffrey P. Blick, "The Archaeology and Ethnohistory of the Dog in Virginia Algonquian Culture as Seen from Weyanoke Old Town," *Papers of the Thirty-First Algonquin Conference* 31 (2000): 6. Regarding the cultural and emotional implications of pre-contact dog burials, see: Darcy Morey, *Dogs: Domestication and the Development of a Social Bond* (New York: Cambridge University Press, 2010); Morey, "Burying Key Evidence," 158–175.

15 Ben F. Koop, "Ancient DNA Evidence of a Separate Origin for North American Indigenous Dogs," in *Dogs through Time: An Archaeological Perspective Proceedings of the 1st ICAZ Symposium on the History of the Domestic Dog*, edited by Susan J. Crockford (Oxford: British Archaeological Reports International Series 889, 2000), 271; Leonard, et al., "Ancient DNA Evidence for Old World Origin of New World Dogs," 1613–1616.

16 Raul Y. Tito, et al. "DNA from Early Holocene American Dog," *American Journal of Physical Anthropology* 145 (2011): 655–656.

17 Tovi M. Anderson, et al., "Molecular and Evolutionary History of Melanism in North American Gray Wolves," *Science* 323 (March 6, 2009): 1339–1346. If one is comfortable with the practice of "upstreaming" (drawing inferences about one historical era based on evidence from another era), then the sources would seem to suggest that black wolves were distributed throughout the South, from Virginia to Florida. Early naturalists consistently noted that the region's wolves were almost always black. Just prior to the American Revolution, the famed naturalist William Bartram observed numerous black wolves while traveling throughout Georgia and Florida. Not long thereafter, French revolutionary Jacquest-Pierre Brissot de Warville recounted seeing black wolves while visiting Thomas Jefferson at Monticello in the Virginia piedmont. Evidence from later centuries reveals much the same. In the 1840s, famed artist and naturalist John James Audubon visited the United States while preparing his manuscript, *The Viviparous Quadrupeds of North America*. Although his artistic rendering of an "American Black Wolf" was presumably based on his observations in the Missouri River valley (plains bison are featured in the background), he elsewhere confirmed that black wolves were prevalent in the South. He had personally observed several black wolf pelts while in North Carolina, and he reported that one-fifth of the wolves recently killed near Colleton, South Carolina, were also black. By 1894, there were several black wolves being housed at the National Zoo in Washington, DC, and the promotional literature informed visitors that "The black wolf ... variety of the American wolf is found principally in the southern part of the United States." Finally, Tappan Gregory provided the earliest photographic evidence in 1935, when he captured a "Florida Black Wolf" on film using a trip-wire camera. William Bartram, *Travels through North and South Carolina, Georgia, East and West Florida* (Philadelphia: James & Johnson, 1791), 172–174; Jacques-Pierre Brissot de Warville, *New Travels in the United States of America* (Bowling Green, OH: Historical Publications Company, 1919), 454–455; John James Audubon and John Bachman, *Viviparous Quadrupeds of North America* vol. 2 (New York: V. G. Audubon, 1851), 130–131; G. W. Orme, "Guide to the National Zoological Park (1894)," 21. Located in "National Zoological Park Publications," Accession Number 12-086, Box 4,

Smithsonian Institution Archives, Washington, DC; Tappan Gregory, "The Black Wolf of the Tensas," *Program of Activities of the Chicago Academy of Sciences* 6 (July, 1935), Record Group 22 – Records of the U. S. Fish & Wildlife Service / Folder – 'Wolf – Ref,' National Archives, College Park, MD.

18 Marcy Norton, "The Chicken or the Iegue: Human–Animal Relationships and the Columbian Exchange," *American Historical Review* 120 (February, 2015): 28–60; Bartolome de Las Casas, *Las Casas on Columbus: a Translation*, edited by Nigel Griffin and Anthony Pagden (Turnhout, Belgium: Brepols Publishers, 1999), 90; Elinor G. K. Melville, *A Plague of Sheep: Environmental Consequences of the Conquest of Mexico* (Cambridge: Cambridge University Press, 1997); John M. Street, "Feral Animals in Hispaniola," *Geographical Review* 52 (July, 1962): 400; Gonzalo Fernandez de Oviedo, *Natural History of the West Indies*, translated and edited by Sterling A. Stoudemire (Chapel Hill: University of North Carolina Press, 1959), 11, 90.

19 Technically speaking, use of the word "Spanish" in this instance is both anachronistic and inaccurate. After all, these expeditions predated the Spanish state by centuries, and the expeditions' participants primarily swore allegiance to the region in which they were born. What is more, all of Europe's earliest expeditions to the South included African participants. Despite these facts, I have retained use of the "Spanish" label in the interest of narrative coherence. Regarding the explorers' regional allegiance, see: Charles Hudson, *Knights of Spain, Warriors of the Sun: Hernando de Soto and the South's Ancient Chiefdoms* (Athens: University of Georgia Press, 1997), 3. Regarding the participation of Africans in these expeditions, see: Jane Landers, *Black Society in Spanish Florida* (Champaign: University of Illinois Press, 1999), 12–13.

20 The standard textbook narrative still credits Ponce de Leon with being the first European to sight mainland North America in 1513. It may be true that the Spanish were slow to realize that a continent lurked nearby, but surely they could not have been *that* slow. It seems incredible that they would travel hundreds of miles across a potentially infinite ocean, and then wait another 21 years before traveling a few extra miles across relatively shallow water to reach North America. It seems especially unlikely given that Indians in the Caribbean were in contact with the mainland and could therefore attest to its existence. Additional evidence is drawn from early maps. The Cantino Planisphere of 1502 clearly shows that mariners had knowledge of some vast landmass that existed northwest of Cuba and Hispaniola. James J. Miller, *An Environmental History of Northeast Florida* (Gainesville: University of Florida Press, 1998), 93; William P. Cummings, *The Southeast in Early Maps*, third edition (Chapel Hill: University of North Carolina Press, 1998), 106; Jerry Brotton, *Trading Territories: Mapping the Early Modern World* (London: Reaktion, 1997), 22–23.

21 When Ponce de Leon tried to establish a colony in Florida in 1521, he brought 200 men and an unspecified number of horses and pigs (as well as cows, sheep, and goats). When Lucas Vázquez de Ayllón tried to land a colony on the Santee River in present-day South Carolina in the summer of 1526, he took more than 600 people, as well as pigs and horses. And when Pánfilo de Narváez tried to establish a colony in 1527, he amassed an army of 300

people, 40 horses, an unspecified number of dogs, but no pigs. Frederick Davis, "Ponce de Leon's Second Voyage and Attempt to Colonize Florida," *Florida Historical Society Quarterly* 14 (July, 1935): 59; Paul E. Hoffman, "Lucas Vázquez de Ayllón's Discovery and Colony," in *The Forgotten Centuries: Indians and Europeans in the American South, 1521–1704*, edited by Charles Hudson and Carmen Chaves Tesser (Athens: University of Georgia Press, 1994), 36–43; Hudson, *Knights of Spain, Warriors of the Sun*, 36–37; Paul E. Hoffman, "Narváez and Cabeza de Vaca in Florida," in *The Forgotten Centuries: Indians and Europeans in the American South, 1521–1704*, edited by Charles Hudson and Carmen Chaves Tesser (Athens: University of Georgia Press, 1994), 53–58.

22 Francis Haines, "Northward Spread of Horses among the Plains Indians," *American Anthropologist* 40, (July–September, 1938): 429–437. Works that credit De Soto's strays with founding the western mustangs Clark Wissler, "The Influence of the Horse in the Development of Plains Culture," *American Anthropologist* 16 (January–March, 1914): 9; "The First Importations of Stock," *Colman's Rural World* 34 (February 3, 1881): 37; Henry W. Herbert, quoted in "Thoroughbred Horse the Highest Type of the Equine Species," *Turf, Field and Stream* 10 (March 1870): 161; "Romantic History of Florida," *DeBow's Review* (March, 1858): 245.

23 There were multiple opportunities for feralization during the expedition. When crossing the Savannah River, numerous pigs were carried downstream and lost from the historical record forever. What is more, Native Americans quickly developed a taste for pork and often tried to steal pigs from the Spaniards. On one occasion, while backtracking across territory they had already covered, the Spaniards discovered that some of the Indians they had left in their wake were now living with pilfered pigs. Finally, it is not entirely clear what happened to all of their hogs by the end of the expedition. Elvas seemed to suggest that the Spaniards killed and salted all of their remaining hogs before heading down the Mississippi River, whereas Garcilaso de la Vega (a sixteenth-century Spaniard who was not part of the De Soto expedition) wrote that Indians stole a canoe loaded with five pigs. Rodrigo Rangel, "Account of the Northern Conquest and Discovery of Hernando de Soto," in *The De Soto Chronicles: the Expedition to North America in 1539–1543*, vol. 1, edited by Lawrence A. Clayton, et al. (Tuscaloosa: University of Alabama Press, 1993), 274; Gentleman of Elvas, "True Relation of the Hardships Suffered by Governor Hernando De Soto & Certain Portuguese Gentlemen during the Discovery of the Province of Florida, now Newly Set Forth by a Gentleman of Elvas," in *The De Soto Chronicles: the Expedition to North America in 1539–1543*, vol. 1, edited by Lawrence A. Clayton, et al. (Tuscaloosa: University of Alabama Press, 1993), 106, 542; Hudson, *Knights of Spain, Warriors of the Sun*, 358; Ralph H. Vigil, "The Expedition of Hernando de Soto and the Spanish Struggle for Justice," in *The Hernando de Soto Expedition: History, Historiography, and 'Discovery' in the Southeast*, edited by Patricia Galloway (Lincoln: University of Nebraska, 2006), 343.

24 When the French explorer Jean Ribault established Fort Caroline on the banks of the modern-day St. Johns River near modern-day Jacksonville in

1564, he reported seeing "wild swyne." Unfortunately, he did not elaborate and subsequent observers did not corroborate his claim. What is more, zooarchaeological evidence suggests that pigs played a relatively small part of Spanish diets in sixteenth-century Florida. Barnet Pavao-Zuckerman and Elizabeth J. Reitz, "Eurasian Domesticated Livestock in Native American Economies," in *Subsistence Economies of Indigenous North American Societies*, edited by Bruce D. Smith (Washington, DC: Smithsonian Institution, 2011), 577–591; Hudson, *Knights of Spain, Warriors of the Sun*, 440; Barnet Pavao-Zuckerman and Elizabeth J. Reitz, "Introduction and Adoption of Animals from Europe," in *Handbook of North American Indians*, vol. 3: *Environment, Origins, and Population*, edited by Douglas H. Ubelaker (Washington, DC: Smithsonian Institution, 2006), 485–491; Jean Ribaut, *The Whole & True Discovery of Terra Florida: A Facsimile Reprint of the London Edition of 1563* (Deland: Florida State Historical Society, 1927), 72–73.

25 More precisely, they took some animals and acquired others. When Richard Grenville and five ships left Plymouth en route for the South in 1585, the convoy contained numerous dogs (mastiffs). When two of the ships were forced into a brief, unexpected layover in Puerto Rico, the prospective colonists appropriated several horses and pigs from among the island's feral droves. Richard Grenville, "Grenville in Virginia," in *Stories from Hakluyt*, edited by Richard Wilson (London: J. M. Dent & Sons, 1921), 171–176.

26 De Soto's army ate untold thousands of native dogs during their trek across the South. Luys Hernández de Biedma, "Relations of the Island of Florida," in *The De Soto Chronicles: the Expedition to North America in 1539–1543*, vol. 1, edited by Lawrence A. Clayton, et al. (Tuscaloosa: University of Alabama Press, 1993), 226, 236; Elvas, "True Relation," 77, 87, 142; Rangel, "Account of the Northern Conquest and Discovery of Hernando de Soto," 281–282.

27 Thomas Harriot, *A brief and true report of the new found land of Virginia*, (New York: Dodd, Mead & Company, [1588], 1903), 53.

28 John Smith, *Lives of Alexander Wilson and Captain John Smith* (Boston: Hilliard, Gray, and Co., [1624] 1834), 378; Gabriel Archer, "A Relatyon written by a gent. of ye Colony," in *Jamestown Voyages Under the First Charter*, edited by Philip L. Barbour (Burlington, VT: Ashgate, 2010), 95–96.

29 George Percy, "Observations by Master George Percy," in *Narratives of Early Virginia*, edited by Lyon Gardner Tyler (New York: Scribner & Sons, [1607] 1959), 8.

30 For reference to the gifted dog, see: John Smith, *Lives of Alexander Wilson and Captain John Smith* (Boston: Hilliard, Gray, and Co., [1624] 1834), 378. Incidentally, Chief Powhatan's given name was actually Wahunsonacock. April Lee Hatfield, "Spanish Colonization Literature, Powhatan Geographies, and English Perceptions of Tsenacommacah," *Journal of Southern History* 69 (May, 2003): 250.

31 Gabriel Archer, "A Relatyon written by a gent. of ye Colony," in *Jamestown Voyages Under the First Charter*, edited by Philip L. Barbour (Burlington, VT: Ashgate, 2010), 95–96.

32 John Smith, *The Generall Historie of Virginia, New-England, and the Summer Isles* (London: Printed by I. D. and I. H. for Michael Sparkes, 1624), 185–186.

33 Smith, *The Generall Historie of Virginia*, 27.
34 Harriot, "A brief and true report of the new found land of Virginia," 53; Christopher Columbus, entry for Tuesday, 16th of October, "Journal of the First Voyage of Columbus," in *The Northmen, Columbus, and Cabot, 985–1503*, edited by Edward Gaylord Bourne (New York: Scribner, 1906), 119-12, 130–132, 142.
35 Smith, *The Generall Historie of Virginia*, 86.
36 Frank E. Grizzard, *The Jamestown Colony: a Political, Social, and Cultural History* (Santa Barbara: ABC-CLIO, 2007), 90.
37 Counseil for Virginia, "A True Declaration of the estate of the Colonie in Virginia," in *Tracts and other Papers Relating Principally to the Origin, Settlement, and Progress of the Colonies in North America from the Discovery of the Country to the Year 1776*, vol. 3, edited by Peter Force (Washington, Wm. Q. Force, 1844), 17.
38 "The Equine FFVs: A Study of the Evidence for the English Horses Imported into Virginia before the Revolution," *The Virginia Magazine of History and Biography* 35 (October, 1927): 329.
39 This was the same hurricane that reportedly inspired William Shakespeare to write *The Tempest*. Lorri Glover and Daniel Blake Smith, *The Shipwreck that Saved Jamestown: The Sea Venture Castaways and the Fate of America* (New York: Henry Holt, 2008).
40 George Percy, "A Trewe Relacyon," reprinted in *Tyler's Quarterly Historical and Genealogical Magazine* 3 (April, 1922): 267. David W. Stahl, et al., "The Lost Colony and Jamestown Droughts," *Science* 280 (1998): 564–567; Dennis Blanton, "Drought as a Factor in the Jamestown Colony, 1607–1612," *Historical Archaeology* 34 (2000): 74–81; Historians continue to debate whether colonists actually resorted to cannibalism during the Starving Time. Rachel B. Herrmann, "The 'tragicall historie': Cannibalism and Abundance in Colonial Jamestown," *William and Mary Quarterly* 68 (January, 2011): 47–74. See also: Nicholas Wade, "Girl's Bones Bear Signs of Cannibalism by Starving Virginia Colonists," *New York Times*, May 1, 2013. Note that the settlers' hardships were exacerbated by the "Little Ice Age." Mark Carey, et al., "Forum: Climate Change and Environmental History," *Environmental History* 19, 2(2014): 281–364; Karen Ordahl Kupperman, "The Puzzle of the American Climate in the Early Colonial Period," *American Historical Review* 87 (December, 1982): 1262–1289; Karen Ordahl Kupperman, *The Jamestown Project* (Cambridge: Harvard University Press, 2007).
41 Virginia DeJohn Anderson, "Somer Islands' 'Hogge Money," *Environmental History* 9 (2004): 128–131.
42 Robert Beverley, *The History and Present State of Virginia* (London, England: R. Parker, 1705), 24–25; Robert Johnson, "The New Life of Virginea: Declaring the former successe and present estate of that plantation (1612)," in *Tracts and other Papers Relating Principally to the Origin, Settlement, and Progress of the Colonies in North America from the Discovery of the Country to the Year 1776*, vol. 1, edited by Peter Force (Washington, DC: Peter Force, 1836), 14.
43 William Strachey, "For the Colony in Virginea Brittannia (1612)," in *Tracts and Other Papers Relating Principally to the Origin, Settlement, and Progress of the Colonies in North America from the Discovery of the Country to the Year 1776*, vol. 3, edited by Peter Force (Washington: Wm. Q. Force, 1844), 15.

44 Beverley, *The History and Present State of Virginia*, 27–28; John Smith, *Travels and Works of Captain John Smith*, part II (New York: Burt Franklin, 1910), 510. It is significant that pigs were generally not allowed to go large in the dense, urbanized parts of England from whence most of the settlers hailed. See: Dolly Jorgensen, "Running Amuck? Urban Swine Management in Late Medieval England," *Agricultural History* (2013): 429–451.
45 Ralph Hamor, *A True Discourse of the Present State of Virginia*, Virginia State Library Publications, no. 3 (Richmond: The Virginia State Library, 1957), 23.
46 "A Note of the Shipping, Men, and Provisions, Sent to Virginia, by the Treasurer and Company in the Yeere 1619," in *The Records of the Virginia Company of London*, vol. 3, edited by Susan Myra Kingsbury (Washington, DC: Government Printing Office, 1933), 118; That same year, John Rolfe wrote that horses had increased in number and become "very vendible." John Rolfe, "A Letter to Sir Edwin Sandys, January 1619/20," *The Records of the Virginia Company of London*, vol. 3, edited by Susan Myra Kingsbury (Washington, DC: Government Printing Office, 1933), 243.
47 Peter Arundle, "Fragment of a Letter to John Smith of Nibley," in *The Records of the Virginia Company of London*, vol. 3, edited by Susan Myra Kingsbury (Washington, DC: Government Printing Office, 1933), 589.
48 "At an Extraordinary Courte Helde for Virginia, July 2, 1621," in *The Records of the Virginia Company of London*, vol. 1, edited by Susan Myra Kingsbury (Washington, DC: Government Printing Office, 1906), 502; For more information on this topic, see: Virginia DeJohn Anderson, "Animals into the Wilderness: The Development of Livestock Husbandry in the Seventeenth-Century Chesapeake," *William and Mary Quarterly* 59 (April, 2002): 377–408.
49 Governor Argall, "Proclamations or Edicts, May 10, 1618," *The Records of the Virginia Company of London*, vol. 3, edited by Susan Myra Kingsbury (Washington, DC: Government Printing Office, 1933), 93.
50 General Assembly of Virginia, "Proceedings of the Virginia Assembly, August 4, 1619," in *Narratives of Early Virginia*, edited by Lyon Gardner Tyler (New York: Scribner & Sons, 1959), 270.
51 Council in Virginia, "A Letter to the Virginia Company of London, April (after 20), 1622, in *The Records of the Virginia Company of London*, vol. 3, edited by Susan Myra Kingsbury (Washington, DC: Government Printing Office, 1933): 612.
52 William Rowlsley, "Letter to his brother, April 3, 1623," in *Records of the Virginia Company*, vol. 2, edited by Susan Myra Kingsbury (Washington, DC: Government Printing Office, 1935), 235.
53 J. Frederick Fausz and John Kukla, "A Letter of Advice to the Governor of Virginia, 1624," *William and Mary Quarterly* 34 (January, 1977): 121.
54 Edward Waterhouse, "A Declaration of the State of the Colony and Affaires in Virginia, 1621," in *The Records of the Virginia Company of London*, vol. 3, edited by Susan Myra Kingsbury (Washington, DC: Government Printing Office, 1933), 557. Meanwhile, at least one settler predicted that colonists would miss the Indians, if only for selfish reasons. John Martin wrote that Indians helped keep the number of wolves down, and he predicted that the colonists would therefore be sorry to see the Indians go. John Martin, "The Manner Howe to Bring the Indians into Subjection, December 15, 1622,"

in *The Records of the Virginia Company of London*, vol. 3, edited by Susan Myra Kingsbury (Washington, DC: Government Printing Office, 1933), 706.

55 Francis Wyatt, "A Proclamation against stealing of beasts and birds of Domesticall & tame nature, September 21, 1623," in *The Records of the Virginia Company of London*, vol. 4, edited by Susan Myra Kingsbury (Washington, DC: Government Printing Office, 1935), 283.

56 Philip Alexander Bruce, *Economic History of Virginia in the Seventeenth Century*, vol. 1 (New York: Macmillan Company, 1907), 312–313.

57 General Assembly of Virginia, "Act VI, March 1629–30," in *Hening's Statutes at Large: being a Collection of all the Laws of Virginia from the First Session of the Legislature*, vol. 1, edited by William Waller Hening (New York: Bartow, 1823), 152.

58 General Assembly of Virginia, "Act LXIII," in *Hening's Statutes at Large: being a Collection of all the Laws of Virginia from the First Session of the Legislature*, vol. 1, edited by William Waller Hening (New York: Bartow, 1823), 176. See also: Earl W. Hayter, "Livestock-Fencing Conflicts in Rural America," *Agricultural History* 37 (January, 1963): 10–20.

59 William Bullock, *Virginia Impartially examined, and left to publick view, to be considered by all Judicious and honest men* (London: John Hammond, 1649).

60 Edward Williams, "Virginia: More Especially the South part thereof, Richly, and Truly Valued," in *Tracts and Other Papers Relating Principally to the Origin, Settlement, and Progress of the Colonies in North America from the Discovery of the Country to the Year 1776*, vol. 3, edited by Peter Force (Washington, DC: Wm. Q. Force, 1844), 14.

61 John Hammond, "Leah and Rachel, or, the Fruitfull Sisters Virginia and Maryland," in *Tracts and Other Papers Relating Principally to the Origin, Settlement, and Progress of the Colonies in North America from the Discovery of the Country to the Year 1776*, vol. 3 edited by Peter Force (Washington, DC: Wm. Q. Force, 1844), 19.

62 Anderson, "Animals into the Wilderness," 398–399; Cary Carson, et al., "New World, Real World: Improvising English Culture in Seventeenth-Century Virginia," *Journal of Southern History* 74 (2008): 31–88.

63 General Assembly of Virginia, "Act CXXV, March, 1661–2," in *Hening's Statutes at Large*, vol. 2, edited by William Waller Hening (New York: Bartow, 1823), 129.

64 William Bullock, "Virginia impartially examined, and left to publick view, to be considered by all Judicious, and honest men" (London: John Hammond, 1649), 8, 51; John Farrer, "A Perfect Description of Virginia," in *Tracts and Other Papers Relating Principally to the Origin, Settlement, and Progress of the Colonies in North America from the Discovery of the Country to the Year 1776*, vol. 2, edited by Peter Force (Washington, DC: William Q. Force, 1838), 3, 39.

65 General Assembly of Virginia, "Act IX, Concerning Stray Horses and Cattell, December 1656," in *Hening's Statutes at Large*, vol. 1, edited by William Waller Hening (New York: Bartow, 1823), 420–421; "Act XVIII, Stray Horses, e&c.," in *Hening's Statutes at Large*, vol. 2, edited by William Waller Hening (New York: Bartow, 1823), 124.

66 General Assembly of Virginia, "Act XLIX, September, 1632," in *Hening's Statutes at Large*, vol. 1, edited by William Waller Hening (New York: Bartow, 1823), 199.

67 General Assembly of Virginia, "Act I, An Introduction to the Acts concerning Indians, March 10, 1655," in *Hening's Statutes at Large*, vol. 2, 395–396.

68 General Assembly of Virginia, "Act XIX, Horses tythable to defray the charges of Wolves heads, December, 1662," in *Hening's Statutes at Large*, vol. 2, edited by William Waller Hening (New York: Bartow, 1823), 178.

69 General Assembly of Virginia, "Act II, Act Repealing the Act Laying a Tax upon Horses, October, 1665," in *Hening's Statutes at Large*, vol. 2, edited by William Waller Hening (New York: Bartow, 1823), 215.
70 General Assembly of Virginia, "Act VI, About Horses, September, 1668," in *Hening's Statutes at Large*, vol. 2, edited by William Waller Hening (New York: Bartow, 1823), 267.
71 General Assembly of Virginia, "Act II, Against Importation of Horses and Mares, October, 1669," in *Hening's Statutes at Large*, vol. 2, edited by William Waller Hening (New York: Bartow, 1823), 271.
72 General Assembly of Virginia, "Act II, Concerning fences, October, 1670," in *Hening's Statutes at Large*, vol. 2, edited by William Waller Hening (New York: Bartow, 1823), 279.
73 Jennings Cropper Wise, "The Chincoteague Pony," *Bit and Spur*, April 1, 1910, 10.
74 William Berkeley, "Enquiries to the Governor of Virginia (1670)," in *Hening's Statutes at Large*, vol. 2, edited by William Waller Hening (New York: Bartow, 1823), 512; Thomas Glover, *An Account of Virginia, its Scituation, Temperature, Productions, Inhabitants and Their Manner of Planting and Ordering Tobacco, &c.*, (Oxford, England: B. H. Blackwell for the Royal Society, 1904), 31.
75 General Assembly of Virginia, "Act XLIX, September, 1632," in *Hening's Statutes at Large*, vol. 1, 199.
76 General Assembly of Virginia, "Act I, March, 1645–6," in *Hening's Statutes at Large*, vol. 1, edited by William Waller Hening (New York: Bartow: 1823), 324–325.
77 General Assembly of Virginia, "Act CXXXVIII, Concerning Indians, March, 1662," in *Hening's Statutes at Large*, vol. 7, edited by William Waller Hening (Richmond: Franklin Press, 1820), 138–139.
78 Edmund Scarburgh, "Propositions Humbly Presented to this Honourable Assembly," in *Hening's Statutes at Large*, vol. 2, edited by William Waller Hening (New York: Bartow, 1823), 202–203.
79 General Assembly of Virginia, "Act VI, An Act commanding such Indians who keep Hoggs to marke the Same, September, 1674," in *Hening's Statutes at Large*, vol. 2, (New York: Bartow, 1823), 316–317.
80 James D. Rice, "Bacon's Rebellion in Indian Country," *Journal of American History* 102 (December, 2014): 726–750; James D. Rice, *Tales from a Revolution: Bacon's Rebellion and the Transformation of Early America* (New York: Oxford University Press, 2012); Brent Tarter, "Bacon's Rebellion, the Grievances of the People, and the Political Culture of Seventeenth-Century Virginia," *Virginia Magazine of History and Biography* 119 (January, 2011): 2–41; Phillip Morgan, *American Slavery, American Freedom: The Ordeal of Colonial America* (New York: Knopf, 1975).
81 His Majesty's Commissioners, "A Narrative of the Rise, Progresse and Cessation of the Late Rebellion in Virginia," in *Samuel Wiseman's Book of Records*, edited by Michael Leroy Oberg (Plymouth: Lexington Books, 2005), 142–143.
82 Thomas J. Wertenbaker (ed.), *Bacon's Rebellion, 1676* (Williamsburg: Virginia 350th Anniversary Celebration Corporation, 1957), 30.
83 Michael Leroy Oberg, "Introduction," *Samuel Wiseman's Book of Record*, edited by Michael Leroy Oberg (Plymouth: Lexington Books, 2005), 18.
84 Michael Leroy Oberg, "Introduction," *Samuel Wiseman's Book of Record*, edited by Michael Leroy Oberg (Plymouth: Lexington Books, 2005), 19; Francis

Moryson, "Questions Proposed by his Majestie and Councill, to Which I returne this Humble plane and true answer," *Samuel Wiseman's Book of Record*, edited by Michael Leroy Oberg (Plymouth: Lexington Books, 2005), 205.
85 General Assembly of Virginia, "Act XIV, An Act for the Further Prevention of Mischief from unrulie horses," June 1676," in *Hening's Statutes at Large*, vol. 2, edited by William Waller Hening (New York: Bartow, 1823), 360–361.
86 Wm. J. Hinke, "Report of the Journey of Francis Louis Michel from Berne, Switzerland, to Virginia, October 2, 1701–December 1, 1702," *Virginia Magazine of History and Biography* 24 (January, 1916): 21–22, 36–37.
87 Beverley, *The History and Present State of Virginia*, 262–263.
88 Beverley, *The History and Present State of Virginia*, 257–258.
89 Wm. J. Hinke, "Report of the Journey of Francis Louis Michel from Berne, Switzerland, to Virginia, October 2, 1701-December 1, 1702," *Virginia Magazine of History and Biography* 24 (January, 1916): 42.
90 Durand of Dauphine, *A Frenchman in Virginia: Being the Memoirs of a Huguenot Refugee in 1686* (Richmond: NP, 1923), 53–54.
91 For more on the food habits of eighteenth-century Virginia colonists, see Lorena S. Walsh, "Feeding the Eighteenth-Century Town Folk, or, Whence the Beef?" *Agricultural History* 73 (1999): 267–280.
92 Justin Roberts and Ian Beamish, "Venturing Out: The Barbadian Diaspora and the Carolina Colony, 1650–1685," in *Creating and Contesting Carolina: Proprietary Era Histories*, edited by Michelle LeMaster and Bradford J. Wood (Chapel Hill: University of North Carolina Press, 2013), 49–72; Alan Taylor, *American Colonies: the Settling of North America* (New York: Penguin, 2001), 224.
93 Max Edelson, *Plantation Enterprise in Colonial South Carolina* (Cambridge: Harvard University Press, 2009), 47; S. Max Edelson, "Clearing Swamps, Harvesting Forests: Trees and the Making of a Plantation Landscape in the Colonial South Carolina Lowcountry," *Agricultural History* 81 (summer, 2007): 390; Mart Stewart, "If John Muir had been an Agrarian: American Environmental History West and South," *Environment and History* 11 (2005): 139–162; Russell R. Menard, "Financing the Lowcountry Export Boom: Capital and Growth in Early South Carolina," *William and Mary Quarterly* 51 (1994): 659–676; Tim Silver, *New Face on the Countryside: Indians, Colonists, and Slaves in South Atlantic Forests, 1500–1800* (New York: Cambridge University Press, 1990), 172; John S. Otto, "Livestock Raising in Early South Carolina, 1670–1700: Prelude to the Rice Plantation Economy," *Agricultural History* 61 (autumn, 1987): 14. See also: Thomas Ashe, *Carolina: or A Description of the Present State of that Country, and the Natural Excellencies Thereof...* (London, England: Privately published, 1682), 20–21; Samuel Wilson, *An Account of the Province of Carolina in America, Together with an Abstract of the Patent, and Several other Necessary and Useful Particulars, to Such as Have Thoughts of Transporting Themselves Thither* (London, England: G. Larkin for Francis Smith, 1682), 14; Maurice Mathews, "A Contemporary View of Carolina in 1680," *The South Carolina Historical Magazine* 55 (July, 1954): 153–159.
94 Otto, "Livestock-Raising in Early South Carolina," 13–24; John Solomon Otto, "The Origins of Cattle-Ranching in Colonial South Carolina, 1670–1715," *South Carolina Historical Magazine* 87 (1986): 117–124.

95 Wilson, *An Account of the Province of Carolina in America*, 14.
96 David J. McCord, *The Statutes at Large of South Carolina*, vol. 2 (Columbia, SC: A. S. Johnston, 1841), 61–63, 72, 106.
97 McCord, *The Statutes at Large of South Carolina*, vol. 2, 81.
98 Regarding the Africans who accompanied Spanish conquistadors, see: Jane Landers, *Black Society in Spanish Florida* (Champaign: University of Illinois Press, 1999): 12–13. Virginia's roughly 3,000 enslaved Africans were owned by a relatively small number (250) of wealthy planters. John C. Coombs, "The Phases of Conversion: A New Chronology for the Rise of Slavery in Early Virginia," *William and Mary Quarterly* 68 (July, 2011): 332–360.
99 Gregory E. O'Malley, "Diversity in the Slave Trade to the Colonial Carolinas," in *Creating and Contesting Carolina: Proprietary Era Histories* (Chapel Hill: University of North Carolina Press, 2013), 236.
100 Jack P. Greene, "Early Modern Southeastern North America and the Broader Atlantic and American Worlds," *Journal of Southern History* 73 (August, 2007): 530–531; W. S. Rossiter, "Population in the Colonial and Continental Periods," *A Century of Population Growth: From the First to the Twelfth Census of the United States, 1790–1900* (Washington, DC: Government Printing Office, 1909), 6–7;
101 Gregory E. O'Malley, "Diversity in the Slave Trade to the Colonial Carolinas," in *Creating and Contesting Carolina: Proprietary Era Histories* edited by Michelle LeMaster and Bradford J. Wood (Chapel Hill: University of North Carolina Press, 2013), 236–239; Philip D. Curtin, *The Atlantic Slave Trade: A Census* (Madison: University of Wisconsin Press, 1972).
102 James L. A. Webb, Jr., "The Horse and Slave Trade between the Western Sahara and Senegambia," *Journal of African History* 34 (1993): 221–246; Silver, *New Face on the Countryside*, 174; Robin Law, *The Horse in West African History: The Role of the Horse in the Societies of Pre-Colonial Africa* (London: International African Institute, 1980); Robin Law, "Horses, Firearms, and Political Power in Pre-Colonial West Africa," *Past & Present* 72 (August, 1976): 112–132.
103 Andrew Sluyter, *Black Ranching Frontiers: African Cattle Herders of the Atlantic World, 1500–1900* (New Haven: Yale University Press, 2012), 136–137; S. Max Edelson, "Nature of Slavery: Environmental Disorder and Slave Agency in Colonial South Carolina," in *Cultures and Identities in Colonial British American*, edited by Robert Olwell and Alan Tully (Baltimore: Johns Hopkins, 2005); 27–28; Gary S. Dunbar, "Colonial Carolina Cowpens," *Agricultural History* 35 (July, 1961): 125–131; Andrew Sluyter, "How Africans and their Descendants Participated in Establishing Open-Range Cattle Ranching in the Americas," *Environment and History* 21 (February, 2015): 77–101; Phillip D. Morgan, *Slave Counterpoint: Black Culture in the Eighteenth-Century Chesapeake and Lowcountry* (Chapel Hill: University of North Carolina Press, 1998), 52; Peter H. Wood, *Black Majority: Negroes in Colonial South Carolina from 1670 through the Stono Rebellion* (New York: Norton, 1974).
104 Thomas J. Little, "The South Carolina Slave Laws Reconsidered, 1670–1700," *South Carolina Historical Magazine* 94 (April, 1993): 86–101; Philip

D. Morgan, "Work and Culture: The Task and the World of Lowcountry Blacks, 1700 to 1880," *William and Mary Quarterly* 39 (1982): 574–575.
105 Ira Berlin, *Many Thousands Gone: The First Two Centuries of Slavery in North America* (Cambridge: Belknap, 1998), 68; Loren Schwininger, "Slave Independence and Enterprise in South Carolina, 1780–1865," *South Carolina Historical Magazine* 93 (April, 1992): 101–103.
106 George Burrington, "Report by George Burrington concerning general conditions in North Carolina, January 1, 1733," in *Colonial Records of North Carolina*, edited by William L. Saunders, vol. 3 (Raleigh, P. M. Hale, 1886), 180–189; Jonathan Edward Barth, "'The Sinke of America': Society in the Antebellum Borderlands of North Carolina, 1663–1729," *North Carolina Historical Review* 87 (January, 2010): 1–27; Noeleen McIlvenna, *A Very Mutinous People: The Struggle for North America, 1660–1713* (Chapel Hill: University of North Carolina Press, 2009), 3, 16, 21–27.
107 Taylor, *American Colonies*, 226.
108 General Assembly of North Carolina, "Chapter XLV, What Fences are Sufficient (1715–6)," in *State Records of North Carolina*, vol. 23, edited by Walter Clark (Goldsboro, NC: Nash Brothers, 1904), 61.
109 Helen C. Rountree, "The Termination and Dispersal of the Nottoway Indians of Virginia," *Virginia Magazine of History and Biography* 95 (April, 1987): 193–194.
110 McIlvenna, *A Very Mutinous People*, 117.
111 North Carolina Governor's Council, "Letter from the North Carolina Governor's Council to the Virginia Governor's Council, June 17, 1707," in *Colonial Records of North Carolina*, vol. 1, edited by William L. Saunders (Raleigh: P. M. Hale, 1886), 659.
112 Edward Hyde, "Letter to Governor Spotswood, January 21, 1710/1711," *Colonial Records of North Carolina*, vol. 1, edited by William L. Saunders (Raleigh: P. M. Hale, 1886), 751.
113 Paul Kelton, *Epidemics and Enslavement: Biological Catastrophe in the Native Southeast* (Lincoln: University of Nebraska, 2007).
114 Christopher Gale, "Letter to his sibling, November 2, 1711," in *Colonial Records of North Carolina*, vol. 1, edited by William L. Saunders (Raleigh: P. M. Hale, 1886), 825–829. See also: William L. Ramsey, *The Yamasee War: A Study of Culture, Economy, and Conflict in the Colonial South* (Lincoln: University of Nebraska Press, 2008), 122.
115 For more on the Tuscarora War: Stephen Feeley, "'Before Long to be Good Friends': Diplomatic Perspectives of the Tuscarora War," in *Creating and Contesting Carolina: Proprietary Era Histories*, edited by Michelle LeMaster and Bradford J. Wood (Chapel Hill: University of North Carolina Press, 2013), 140–163; James Taylor Carson, "Histories of the 'Tuscarora War,'" in *Creating and Contesting Carolina: Proprietary Era Histories*, edited by Michelle LeMaster and Bradford J. Wood (Chapel Hill: University of North Carolina Press, 2013), 186–210; Alan Gallay, *Indian Slave Trade: The Rise of English Empire in the American South, 1670–1717* (New Haven: Yale University Press, 2002), 259–287.

116 Benjamin Simson, et al., "Petition from the Inhabitants of the Neuse River area concerning attacks by Native Americans," in *Colonial Records of North Carolina*, vol. 1, edited by William L. Saunders (Raleigh: P. M. Hale, 1886), 819–820.
117 Thomas Pollock,"Letter from Thomas Pollock to the Lords Proprietor of Carolina," in *Colonial Records of North Carolina*, vol. 1, edited by William L. Saunders (Raleigh: P. M. Hale, 1886), 873.
118 North Carolina Governor's Council, "Minutes of the North Carolina Governor's Council, August 10/11, 1714," in *Colonial Records of North Carolina*, vol. 2, edited by William L. Saunders (Raleigh: P. M. Hale, 1886), 141. See also: Michelle LeMaster, "In the 'Scolding House': Indians and the Law in Eastern North Carolina, 1684–1760," *North Carolina Historical Review* 83 (April, 2006): 225.
119 Silver, *A New Face on the Countryside*, 179; Richard Haan, "The 'Trade Do's Not Flourish as Formerly': the Ecological Origins of the Yamassee War," *Ethnohistory* 28 (Autumn, 1981): 341; Mart Stewart, *What Nature Suffers to Groe: Life, Labor, and Landscape on the Georgia Coast, 1680–1920* (Athens: University of Georgia Press, 2002), 27.
120 David La Vere, *The Tuscarora War: Indians, Settlers, and the Fight for the Carolina Colonies* (Chapel Hill: University of North Carolina Press, 2013); Robbie Ethridge, *From Chicaza to Chicasaw: The European Invasion and the Transformation of the Mississippian World, 1540–1715* (Chapel Hill: University of North Carolina Press, 2010): 194–231, 237; Steven J. Oatis, *A Colonial Complex: South Carolina's Frontiers in the Era of the Yamasee War, 1680–1730* (Lincoln: University of Nebraska Press, 2004). According to Gallay, slavers captured 30,000 to 50,000 Native Americans before 1715. Gallay, *Indian Slave Trade*, 299.
121 North Carolina Governor's Council, "Minutes of the North Carolina Governor's Council, June 14, 1722," in *Colonial Records of North Carolina*, vol. 2, edited by William L. Saunders (Raleigh: P. M. Hale, 1886), 457–458;
122 Denise I. Bossy, "Indian Slavery in Southeastern Indian and British Societies, 1670–1730," in *Indian Slavery in Colonial America*, edited by Alan Gallay (Lincoln: University of Nebraska Press, 2009), 219.
123 John Lawson, *A New Voyage to Carolina* (London: s.n. 1709), 8, 164. Some colonists planted entire orchards for free-ranging swine. For examples, see: James Glen, *A Description of South Carolina, Containing Many Curious and Interesting Particulars Relating to the Civil, Natural, and Commercial History of that Colony* (London: R. & J. Dodsley, 1761), 68; John Hairr, "John Lawson's Observations on the Animals of Carolina," *North Carolina Historical Review* 88 (July, 2011): 312–332. On the surprisingly long history of barbecue in the South, see: John Shelton Reed, "There's a Word for it – The Origins of 'Barbecue,'" *Southern Cultures* 13 (2007): 138–146.
124 Mark Catesby, *The Natural History of Carolina, Florida and the Bahama Islands: Containing the Figures of Birds, Beasts, Fishes, Serpents, Insects, and Plants*, vol. II (London: Mark Catesby, 1754), xxxie; E. Charles Nelson and David J. Elliott, *The Curious Mister Catesby: A 'Truly Ingenious' Naturalist Explores New Worlds* (Athens: University of Georgia Press, 2015).

125 William Byrd, *History of the Dividing Line and Other Tracts*, vol. 1 (Richmond: N. P., 1866 [1728]), 32–33. For more on Byrd, see: Kevin Joel Berland (ed.), *The Dividing Line Histories of William Byrd II of Westover* (Chapel Hill: University of North Carolina Press, 2013).
126 Richard Everard, "Letter to the Board of Trade of Great Britain," in *Colonial Records of North Carolina*, vol. 2, edited by William L. Saunders (Raleigh: P. M. Hale, 1886), 761.
127 General Assembly of North Carolina, "Minutes of the Lower House of the North Carolina General Assembly, November 5–8, 1733," in *Colonial Records of North Carolina*, vol. 3, edited by William L. Saunders (Raleigh: P. M. Hale, 1886), 621; Jack Temple Kirby, "Virginia's Environmental History: A Prospectus," *Virginia Magazine of History and Biography* 99 (October, 1991): 475.
128 James Westfall Thompson, *A History of Livestock Raising in the United States, 1607–1860* (Washington, DC: United States Department of Agriculture, 1942), 64.
129 Christopher B. Rodning, "Reconstructing the Coalescence of Cherokee Communities in Southern Appalachia," in *The Transformation of the Southeastern Indians, 1540–1760*, edited by Robbie Ethridge and Charles Hudson (Jackson: University of Mississippi Press, 2008), 155–176; Steven Hahn, *The Invention of the Creek Nation, 1670–1763* (Lincoln: University of Nebraska Press, 2004); Tom Hatley, *The Dividing Paths: Cherokees and South Carolinians through the Revolutionary Era* (New York: Oxford University Press, 1993), 38–39. According to Joshua Piker, the Creek obtained horses by raiding Spanish missions in western Florida. It is also possible that they obtained horses from the nearby Choctaw Indians, who lived in modern-day Alabama and Mississippi. According to James Taylor Carson, the Choctaw got their horses from Plains Indians to the west. Joshua Piker, *Okfuskee: A Creek Indian Town in Colonial America* (Cambridge: Harvard University Press, 2009), 119–120; James Taylor Carson, "Horses and the Economy and Culture of the Choctaw Indians, 1690–1840," *Ethnohistory* 42 (summer, 1995): 496.
130 John Brickell, *The Natural History of North-Carolina* (Raleigh: North Carolina Public Libraries, 1911), 52–54.
131 McCord, *The Statutes at Large of South Carolina*, vol. 2, 261; General Assembly of South Carolina, "An Act to prevent stealing of Horses and Neat Cattle," in *The Statutes at Large of South Carolina*, vol. 2, edited by Thomas Cooper (Columbia, SC: A. S. Johnson, 1837), 261.
132 August Gottlieb Spangenberg, "Diary during journey to North Carolina, September 1752," in *Colonial Records of North Carolina*, vol. 4, edited by William L. Saunders (Raleigh: P. M. Hale, 1886), 1312.
133 General Assembly of Virginia, "Act VIII, An act for the better improveing the breed of Horses, October 1686," in *Hening's Statutes at Large*, vol. 3, edited by William Waller Hening (Philadelphia: Thomas DeSilver, 1823), 35–37.
134 "Chap. VIII, An act to restrain the keeping too great a number of Horses and Mares, and for amending the breed, October 1713," in *Hening's Statutes at Large*, vol. 4, 46–49; "Chap. XL, An Act to restrain the keeping to great a

number of horses and mares, and for amending the breed, October 1748," in *Hening's Statutes at Large*, vol. 6, 118–120.
135 General Assembly of North Carolina, "An Act to restrain the keeping too great a Number of Horses and Mares, and for amending the Breed, November 23, 1723," in *State Records of North Carolina*, vol. 23, edited by Walter Clark (Goldsboro, NC: Nash Brothers, 1904), 103–110; General Assembly of North Carolina, "Amends/Augments act to improve breed of horses, November 1768," in *State Records of North Carolina*, vol. 23, edited by Walter Clark (Goldsboro, NC: Nash Brothers, 1904), 759–783.
136 Quoted in: W. G. S. "Racing in Colonial Virginia," *Virginia Magazine of History and Biography* 2 (January, 1895): 299–300.
137 Randy J. Sparks, "Gentleman's Sport: Horse Racing in Antebellum Charleston," *South Carolina Historical Magazine* 93 (January, 1992): 17, 31.
138 Kenneth Cohen, "Well Calculated for the Farmer: Thoroughbreds in the Early National Chesapeake, 1790–1850," *Virginia Magazine of History and Biography* 115 (2007): 378.
139 Brickell, *The Natural History of North-Carolina*, 39.
140 Quoted in W. G. S., "Racing in Colonial Virginia," 299–230.
141 Nancy L. Struna, "The Formalizing of Sport and the Formation of an Elite: The Chesapeake Gentry, 1650–1720," *Journal of Sport History* 13 (1986): 212–234; T. H. Breen, "Horses and Gentlemen: The Cultural Significance of Gambling among the Gentry of Virginia," *William and Mary Quarterly* 34 (April, 1977): 239–257.
142 Edward Williams, "Virginia: More especially the South part thereof, Richly and truly valued, 1650," in *Tracts and other Papers Relating Principally to the Origin, Settlement, and Progress of the Colonies in North America from the Discovery of the Country to the Year 1776*, vol. 3 (Washington, DC: Wm. Q. Force, 1844), 37.
143 Quoted in Sarah Hand Mecham, "Pets, Status, and Slavery in the Late-Eighteenth-Century Chesapeake," *Journal of Southern History* 77 (August, 2011): 550.
144 "Treaty between North Carolina and King Hagler and the Catawba Indians," in *Colonial Records of North Carolina*, vol. 5, edited by William L. Saunders (Raleigh: Josephus Daniels, 1887), 141–144. For more on the Catawbas, see James Merrell, *The Indians' New World: Catawbas and Their Neighbors from European Contact through the Era of Removal* (Chapel Hill: University of North Carolina Press, 1989).
145 Quoted in Robert Weir, *Colonial South Carolina: A History* (Columbia: University of South Carolina Press, 1997), 212.
146 General Assembly of North Carolina, "Chapter V, An additional Act, for appointing Toll-Books, and for preventing People from driving Horses, Cattle, or Hogs, to other Persons' Lands," *State Records of North Carolina*, vol. 23, edited by Walter Clark (Goldsboro, NC: Nash Brothers, 1904), 113.
147 General Assembly of Virginia, "Chap. XLII, An Act for preserving the breed of Sheep, February 1752," in *Hening's Statues at Large*, vol. 6, edited by William Waller Hening (Richmond: W. W. Gray, 1819), 295–296.
148 Lawson, *A New Voyage to Carolina*, 38.

149 Catesby, *The Natural History of Carolina*, xxvie.
150 Beverley, *The History and Present State of Virginia*, 254–255; Brickell, *The Natural History of North-Carolina*, 331–332.
151 Stewart, *What Nature Suffers to Groe*, 39; Mart A. Stewart, "'Whether Wast, Deodand, or Stray': Cattle, Culture, and the Environment in Early Georgia," *Agricultural History* 65 (1991): 1.
152 Stewart, *What Nature Suffers to Groe*, 28; Ethridge, *From Chicaza to Chicasaw*, 244; John T. Juricek, *Colonial Georgia and the Creeks: Anglo-Indian Diplomacy on the Southern Frontier* (Gainesville: University of Florida Press, 2010), 3.
153 Stewart, *What Nature Suffers to Groe*, 47; Juricek, *Colonial Georgia and the Creeks*, 57–58.
154 Stewart, *What Nature Suffers to Groe*, 55; Stewart, "'Whether Wast, Deodand, or Stray,'" 1–28.
155 General Oglethorpe, Letter to the Trustees, July 4, 1739," in *Colonial Records of the State of Georgia*, vol. 22, edited by Allen D. Candler (Atlanta: Chas. P. Byrd, 1913), 169.
156 Wm. Ewen, "Letter to the Honble Trustees, December 4, 1740," in *Colonial Records of the State of Georgia*, vol. 22, edited by Allen D. Candler (Atlanta: Chas. P. Byrd, 1913), 457.
157 Samuel Perkins, "Letter to the Trustees, May 4, 1741," in *Colonial Records of the State of Georgia*, vol. 23, edited by Allen D. Candler (Atlanta: Chas. P. Byrd, 1914), 27.
158 Mr. Thos. Upton, "Letter to the Honble Trustees, September 2, 1743," in *Colonial Records of the State of Virginia*, vol. 23, edited by Allen D. Candler (Atlanta: Chas. P. Byrd, 1914), 163.
159 Mr. Stephens, "Letter to Mr. Benj. Martin, September 15, 1743," in *Colonial Records of the State of Georgia*, vol. 24, edited by Allen D. Candler, revised by Lucian Lamar Knight (Atlanta: Chas. P. Byrd, 1915), 98.
160 Quoted in Drew A. Swanson, *Remaking Wormsloe Plantation: The Environmental History of a Lowcountry Landscape* (Athens: University of Georgia Press, 2012), 36.
161 Mary Bullard, *Cumberland Island: a History* (Athens: University of Georgia Press, 2003), 29.
162 Thomas Jones, "Letter to the Trustees of the State of Georgia," *Colonial Records of the State of Georgia*, vol. 23, edited by Allen D. Candler (Atlanta: Chas. P. Byrd, 1914), 492.
163 John Pyre, "Letter to the Trustees Accountant, August 27, 1746," *Colonial Records of the State of Georgia*, vol. 25, edited by Allen D. Candler, revised by Lucian Lamar Knight (Atlanta: Chas. P. Byrd, 1915), 97.
164 Mr. Stephens', "Letter to Mr. Benj. Martin, September 15, 1743," in *Colonial Records of the State of Georgia*, vol. 24, edited by Allen D. Candler, revised by Lucian Lamar Knight (Atlanta: Chas. P. Byrd, 1915), 98; Henry Parker, James Habersham, and N. Jones, "Letter from the President, & Assistants, to the Trustees Accountant, August 8, 1751," in *Colonial Records of the State of Georgia*, vol. 26, edited by Allen D. Candler, revised by Lucian Lamar Knight (Atlanta: Chas. P. Byrd, 1916), 263–265.

165 William Stephen, et al., "Letter from the President and Assistants to Benjamin Martyn Esqu., July 25, 1750," in *Colonial Records of the State of Georgia*, vol. 26, edited by Allen D. Candler, revised by Lucian Lamar Knight (Atlanta: Chas. P. Byrd, 1916), 38–39.

166 James Van Horn Melton, "From Alpine Miner to Low-Country Yeoman: The Transatlantic Worlds of a Georgia Salzburger," *Past & Present* 201 (2008): 97–140; Stewart, *What Nature Suffers to Groe*, 56; Stewart, "'Whether Wast, Deodand, or Stray,'" 8.

167 Stewart, *What Nature Suffers to Groe*, 75.

168 John Martin Bolzius, "Letter to Mr. Verest," February 22, 1744/5," in *Colonial Records of the State of Georgia*, vol. 24, edited by Allen D. Candler, revised by Lucian Lamar Knight (Atlanta: Chas. P. Byrd, 1915), 358–362.

169 John Bolzius, "Letter to the Honble. James Vernon Esqu., January 5, 1747–48," in *Colonial Records of the State of Georgia*, vol. 25, edited by Allen D. Candler, revised by Lucian Lamar Knight (Atlanta: Chas. P. Byrd, 1915), 254.

170 John Martin Bolzius, "Letter to Mr. Verest," February 22, 1744/5," in *Colonial Records of the State of Georgia*, vol. 24, edited by Allen D. Candler, revised by Lucian Lamar Knight (Atlanta: Chas. P. Byrd, 1915), 358–362.

171 Henry Parker, et al., "Letter from the Vice President and Assistants to the Secretary, January 2, 1750," in *Colonial Records of the State of Georgia*, vol. 24, edited by Allen D. Candler, revised by Lucian Lamar Knight (Atlanta: Chas. P. Byrd, 1915), 115–117.

172 Noeleen McIlvenna, *The Short Life of Free Georgia: Class and Slavery in the Colonial South* (Chapel Hill: University of North Carolina Press, 2015); Taylor, *American Colonies*, 243; Watson W. Jennings, *Cultivating Race: The Expansion of Slavery in Georgia, 1750–1860* (Lexington: University Press of Kentucky, 2012).

173 General Assembly of Georgia, "An Act to regulate Fences in the Province of Georgia, March 7, 1755," in *Colonial Records of the State of Georgia*, vol. 18, edited by Allen D. Candler (Atlanta: Chas. P. Byrd, 1910), 73–74.

174 General Assembly of Georgia, "An act for the better regulating fences in the province of Georgia, March 27, 1759," in *Colonial Records of the State of Georgia*, vol. 18, edited by Allen D. Candler (Atlanta: Chas. P. Byrd, 1910), 304.

175 George Fenwick, "A German Surgeon on the Flora and Fauna of Colonial Georgia: Four Letters of Johann Christoph Bornemann, 1753–1755," *The Georgia Historical Quarterly* 76 (winter, 1992): 891–914.

176 Kathryn E. Holland, "The Creek Indians, Blacks, and Slavery," *Journal of Southern History* 57 (November, 1991): 605; Carson, "Horses and the Economy and Culture of the Choctaw Indians, 1690–1840," 497.

177 General Assembly of Georgia, "An Act to Prevent Stealing of Horses and Neat Cattle And for the more Effectual Discovery and Punishment of such Persons as shall unlawfully brand, mark, or kill the same, March 27, 1759," *Colonial Records of the State of Georgia*, vol. 18, edited by Allen D. Candler (Atlanta: Chas. P. Byrd, 1910), 319–327; The Commons House of Assembly, "A Memorial of John Francis Williams, February, 1768," *Colonial Records of the State of Georgia*, vol. 14, edited by Allen D. Candler (Atlanta: Franklin-Turner, 1907), 547; William Stephen, et al, "Letter from the President and Assistants to Benjamin

Martyn, Esqu., July 25, 1750," *Colonial Records of the State of Georgia*, vol. 26, edited by Allen D. Candler, revised by Lucian Lamar Knight (Atlanta: Chas. P. Byrd, 1916), 38–39. For other examples, see: Joshua Piker, "Colonists and Creeks: Rethinking the Pre-Revolutionary Southern Backcountry," *Journal of Southern History* 70 (August, 2004): 515–516; Governor and Council, "Petitions of Malefactors under Sentence of Death, July 29, 1767," *Colonial Records of the State of Georgia*, vol. 10, edited by Allen D. Candler (Atlanta: Franklin-Turner, 1907), 245; Governor and Council, "Governor's Answer to Indian's second Talk, September 6, 1768," *Colonial Records of the State of Georgia*, vol. 10, edited by Allen D. Candler (Atlanta: Franklin-Turner, 1907), 585; John L. Nichols, "Alexander Cameron, British Agent among the Cherokee, 1764–1781," *South Carolina Historical Magazine* 97 (April, 1996): 99.
178 Barnet Pavao-Zuckerman, "Deerskins and Domesticates: Creek Subsistence and Economic Strategies in the Historic Period," *American Antiquity* 72 (January, 2007): 5–33.
179 Bernard Romans, *A Concise Natural History of East and West Florida* (New York: R. Aitken, 1776), 84, 93, 97.
180 General Assembly of Georgia, "An Act to Amend an Act for the better Regulating the Town of Savannah and for Ascertaining the Common thereunto belonging and also to Authorise and Empower the Church Wardens and Vestry of the Parish of Christ Church to Appoint a Beadle for Purposes herein mentioned March 6, 1766," in *Colonial Records of the State of Georgia*, vol. 18, edited by Allen D. Candler (Atlanta: Chas. P. Byrd, 1910), 758.
181 Paul E. Hoffman, *Florida's Frontiers* (Bloomington: Indiana University Press, 2002), 209.
182 Elizabeth Reitz, "Vertebrate Fauna from Seventeenth-Century St. Augustine," *Southeastern Archaeology* 11 (winter, 1992): 79–94.
183 Hoffman, *Florida's Frontiers*, 174, 202, 204, 209, 216; Edward E. Baptist, *Creating an Old South: Middle Florida's Plantation Frontier before the Civil War* (Chapel Hill: University of North Carolina Press, 2002), 11.
184 Kevin Kokomoor, "A Re-Assessment of Seminoles, Africans, and Slavery on the Florida Frontier," *Florida Historical Quarterly* 88 (fall, 2009): 209–236.
185 Mark V. Barrow, "William Bartram," in *New Encyclopedia of Southern Culture*, vol. 8: *Environment*, edited by Martin Melosi (Chapel Hill: University of North Carolina Press, 2007): 191–194.
186 William Bartram, *Travels through North and South Carolina, Georgia, East and West Florida* (Philadelphia: James & Johnson, 1791), 185–186; Kathryn E. Holland Braund, *Deerskins and Duffels: The Creek Indian Trade with Anglo-America* (Lincoln: University of Nebraska Press, 1993), 176.
187 Bartram, *Travels*, 180, 194–195, 208.
188 Bartram, *Travels*, 53.
189 Bartram, *Travels*, 190, 232.
190 Christopher Camuto, *Another Country: Journeying toward the Cherokee Mountains* (New York: Henry Holt and Company, 1997), 50.
191 Bartram, *Travels*, 172–174. For more on the wolf's aural legacy, see: Peter A. Coates, "The Strange Stillness of the Past: Toward an Environmental History of Sound and Noise," *Environmental History* 10 (October, 2005): 636–665.

192  Camuto, *Another Country*, 50.
193  Durand of Dauphine, *A Frenchman in Virginia*, 112–114.
194  Hugh Jones, *The Present State of Virginia* (New York: J. Sabin, 1865 [1724]), 51.
195  Jack Temple Kirby, "Virginia's Environmental History: A Prospectus," *Virginia Magazine of History and Biography* 99 (October, 1991): 457.
196  Paul M. Pressly, "Scottish Merchants and the Shaping of Colonial Georgia," *Georgia Historical Quarterly* 91 (2007): 135–168; Warren R. Hofstra, *The Planting of New Virginia: Settlement and Landscape in the Shenandoah Valley* (Baltimore: Johns Hopkins University Press, 2004); David T. Gleeson, *The Irish in the South, 1815–1877* (Chapel Hill: University of North Carolina Press, 2001); Henry A. Gemery, "The White Population of the Colonial United States, 1607–1790," in *A Population History of North America*, edited by Michael R. Haines and Richard H. Steckel (New York: Cambridge University Press, 2000), 143–190; Silver, *New Face on the Countryside*, 1.
197  Russell Thornton, "Population History of Native North Americans," in *A Population History of North America*, edited by Michael R. Haines and Richard H. Steckel (New York: Cambridge University Press, 2000), 9–50; Michael R. Haines and Richard H. Steckel, *A Population History of North America* (New York: Cambridge University Press, 2000).
198  Peter H. Wood, "The Changing Population of the Colonial South: An Overview by Race and Region, 1685–1790," in *Powhatan's Mantle: Indians in the Colonial Southeast*, edited by Gregory A. Waselkov, et al. (Lincoln: University of Nebraska Press, 2006), 60.
199  Benjamin Madley, "Reexamining the American Genocide Debate: Meaning, Historiography, and New Methods," *America Historical Review* 120 (February, 2015): 98–139; Matthew Jennings, *New Worlds of Violence: Cultures and Conquests in the Early American Southeast* (Knoxville: University of Tennessee Press, 2011); April Lee Hatfield, "Colonial Southeastern Indian History," *Journal of Southern History* 73 (August, 2007): 570; Paul Kelton, "The Great Southeastern Smallpox Epidemic, 1696–1700: The Region's First Major Epidemic?" in *The Transformation of the Southeastern Indians, 1540–1760*, edited by Robbie Ethridge and Charles Hudson (Oxford: University of Mississippi Press, 2002), 21–37.
200  Marvin T. Smith, "Aboriginal Population Movements in the Postcontact Southeast," in *The Transformation of the Southeastern Indians, 1540–1760*, edited by Robbie Ethridge and Charles Hudson (Oxford: University of Mississippi Press, 2008), 3–20; Helen C. Rountree, "Trouble Coming Southward: Emanations through and from Virginia, 1607–1675," in *The Transformation of the Southeastern Indians, 1540–1760*, edited by Robbie Ethridge and Charles Hudson (Oxford: University of Mississippi Press, 2008), 65–78; Wood, "The Changing Population of the Colonial South," 60.
201  Wood, "The Changing Population of the Colonial South," 60; Piker, *Okfuskee*, 98.
202  Morgan, *Slave Counterpoint*, 57.
203  Gregory E. O'Malley, *Final Passages: The Intercolonial Slave Trade of British America, 1619–1807* (Chapel Hill: University of North Carolina Press,

2014); Gregory E. O'Malley, "Diversity in the Slave Trade to the Colonial Carolinas," in *Creating and Contesting Carolina: Proprietary Era Histories*, edited by Michelle LeMaster and Bradford J. Wood (Chapel Hill: University of North Carolina Press, 2013), 245; Paul M. Pressly, *On the Rim of the Caribbean: Colonial Georgia and the British Atlantic World* (Athens: University of Georgia Press, 2013).
204 Wood, "The Changing Population of the Colonial South," 60; Lorena S. Walsh, "The African American Population of the Colonial United States," in *A Population History of North America*, edited by Michael R. Haines and Richard H. Steckel (New York: Cambridge University Press, 2000), 191–240.
205 Frederick Douglass Opie, *Hog and Hominy: Soul Food from Africa to America* (New York: Columbia University Press, 2010), 29; Morgan, *Slave Counterpoint*, 375; Philip D. Morgan, "Work and Culture: The Task and the World of Lowcountry Blacks, 1700 to 1880," *William and Mary Quarterly* 39 (1982): 574–575.

## 4 Nascent Domestication Initiatives and their Effects on Ferality

1 Peter H. Wood, "The Changing Population of the Colonial South: An Overview by Race and Region, 1685–1790," in *Powhatan's Mantle: Indians in the Colonial Southeast*, edited by Gregory A. Waselkov, et al. (Lincoln: University of Nebraska Press, 2006), 60.
2 Benjamin Smith Barton, "On Indian Dogs," *The Philosophical Magazine* 15 (February, 1803): 1–9, 136–143.
3 John Smith, *The Generall Historie of Virginia, New-England, and the Summer Isles* (London: Printed by I. D. and I. H. for Michael Sparkes, 1624), 27.
4 Barton, "On Indians Dogs," 6, 136. Regarding the purportedly admixed nature of Native American peoples, see: Theda Purdue, *Mixed Blood Indians: Racial Construction in the Early South* (Athens: University of Georgia Press, 2010); Claudio Saunt, et al., "Rethinking Race and Culture in the Early South," *Ethnohistory* 53 (2006): 399–405; Andrew Frank, *Creeks and Southerners: Biculturalism on the Early American Frontier* (Lincoln: University of Nebraska Press, 2005); Theda Purdue, "Race and Culture: Writing the Ethnohistory of the Early South," *Ethnohistory* 51 (2004): 701–723.
5 Ethan Moore, "From Sikwa to Swine: The Hog in Cherokee Culture and Society, 1750–1840," *Native South* 4 (2011): 105–120. See also: Douglas C. Wilms, "Agrarian Progress in the Cherokee Nation Prior to Removal," in *Essays on the Human Geography of the Southeastern United States*, 16 (Carrollton: West Georgia College, 1977), 1–15; Tom Hatley, *The Dividing Paths: Cherokees and South Carolinians through the Revolutionary Era* (New York: Oxford University Press, 1993), 61–63.
6 Kevin Kokomoor, "Creeks, Federalists, and the Idea of Coexistence in the Early Republic," *Journal of Southern History* 81 (November, 2015): 803–842;
7 Barnet Pavao-Zuckerman, "Deerskins and Domesticates: Creek Subsistence and Economic Strategies in the Historic Period," *American Antiquity* 72

(January, 2007): 5–33; Claudio Saunt, et al., "Rethinking Race and Culture in the Early South," *Ethnohistory* 53 (2006): 160–161, 171.
8 Joshua Piker, "Colonists and Creeks: Rethinking the Pre-Revolutionary Southern Backcountry," *Journal of Southern History* 70 (August, 2004): 503–540. The Choctaw who lived to the west, in present-day Mississippi, followed a very similar pattern when it came to the slow adoption of pigs. Richard White, *Roots of Dependency: Subsistence, Environment, and Social Change among the Choctaws, Pawnees, and Navajos* (Lincoln: University of Nebraska Press, 1988).
9 Edward E. Baptist, "The Migration of Planters to Antebellum Florida: Kinship and Power," *Journal of Southern History* 62 (August, 1996): 527–554.
10 Kevin Kokomoor, "A Re-Assessment of Seminoles, Africans, and Slavery on the Florida Frontier," *Florida Historical Quarterly* 88 (fall, 2009): 213. See also: Nathaniel Millett, "Defining Freedom in the Atlantic Borderlands of the Revolutionary Southeast," *Early American Studies* 5 (2007): 367–394.
11 Theda Purdue and Michael Green, *The Cherokee Nation and the Trail of Tears* (N ew York: VikingPenguin, 2007).
12 Quoted in James W. Covington, "Cuban Bloodhounds and the Seminoles," *Florida Historical Quarterly* 33 (October, 1954): 14. See also: Theda Purdue, "Legacy of Indian Removal," *Journal of Southern History* 78 (2012): 3–36; Sara E. Johnson, "'You Should Give them Blacks to Eat:' Waging Inter-American Wars of Torture and Terror," *American Quarterly* 61 (March, 2009): 79–80; Edward E. Baptist, *Creating an Old South: Middle Florida's Plantation Frontier before the Civil War* (Chapel Hill: University of North Carolina Press, 2002): 157; Greg O'Brien, "The Conqueror Meets the Unconquered: Negotiating Cultural Boundaries on the Post-Revolutionary Southern Frontier," *Journal of Southern History* 67 (February, 2001): 39–72; "Florida Blood Hounds," *Burlington Free Press*, March 6, 1840, 2; E. W. S., "The Bloodhound War," *Liberator* 10 (January 24, 1840), 16, 47.
13 Henry C. Wright, "The Bloodhound Party – The Bloodhound Candidate," *Liberator* (July 7, 1848): 107; "General Taylor and the Bloodhounds," *Liberator* 18 (March 3, 1848): 1.
14 Examples from the antebellum period include: William J. Anderson, *Life and Narrative of William J. Anderson: Twenty Four Years a Slave* (Chicago: Daily Tribune Book and Job Printing Office, 1857), 48; "The Bloodhounds of Slavery," *Liberator* 25 (March 23, 1855): 48; James Curry, "Narrative of James Curry, A Fugitive Slave," *Liberator* 10 (January 10, 1840): 1. Meanwhile, historian Walter Johnson describes the relationship between slaves and slave-hunting dogs in great detail in his recent book: Walter Johnson, *River of Dark Dreams: Slavery and Empire in the Cotton Kingdom* (Cambridge: Harvard University Press, 2013): 234–243.
15 Rev. Horace Moulton, "Narrative and Testimony," in *American Slavery as It Is: Testimony of a Thousand Witnesses*, edited by Theodore Dwight Weld (New York: American Anti-Slavery Society, 1839), 21.
16 Frederick Douglass, *My Bondage and My Freedom* (New York: Orton and Mulligan, 1855).
17 Quoted in John Campbell, "The Seminoles, the 'Bloodhound War,' and Abolitionism, 1796–1865," *Journal of Southern History* 72 (May, 2006): 259.

18 Thomas Wentworth Higginson, "Memoir of Thomas Wentworth Higginson," in *Army Life in a Black Regiment* (Boston: Fields, Osgood & Co., 1870), 296.
19 Frederick Law Olmsted, *A Journey in the Seaboard Slave States: With Remarks on their Economy* (New York: Dix and Edwards, 1856), 159. Harriett Beecher Stowe relied on Olmsted's accounts when describing slave-hunting dogs in *Dred: A Tale of the Great Dismal Swamp* (Boston: Phillips, Sampson, and Co., 1856). In the 1830s, twenty years prior to Olmsted's visit, famed slave rebel Nat Turner hid in the Dismal Swamp following his failed insurrection until he was discovered by (what else?) a dog.
20 Olmsted, *A Journey in the Seaboard Slave States*, 160–161, 163. See also: John Hope Franklin and Loren Schweninger, *Runaway Slaves: Rebels on the Plantation* (New York: Oxford University Press, 2000), 160–164.
21 Thomas G. Andrews, "Beasts of the Southern Wild: Slaveholders, Slaves, and Other Animals in Charles Ball's Slavery in the United States," in *Rendering Nature: Animals, Bodies, Places, and Politics*, edited by Marguerite S. Shaffer and Phoebe S. K. Young (Philadelphia: University of Pennsylvania Press, 2015): 21–47; John Campbell, "'My Constant Companion': Slaves and their Dogs in the Antebellum South," in *Working Toward Freedom: Slave Society and Domestic Economy in the American South*, edited by Larry E. Hudson (Rochester: University of Rochester Press, 1994), 53–76.
22 Campbell, "'My Constant Companion,'" 56.
23 Kym S. Rice, "Dogs," in *World of a Slave: Encyclopedia of the Material Life of Slaves in the United States*, edited by Kym S. Rice and Martha B. Katz-Hyman (Santa Barbara: ABC-CLIO, 2011), 181. See also: Kathleen M. Hilliard, *Masters, Slaves, and Exchange: Power's Purchase in the Old South* (New York: Cambridge University Press, 2013).
24 Campbell, "'My Constant Companion,'" 55, 57, 59, 60; Andrews, "Beasts of the Southern Wild," 38–39, 41.
25 George Washington, "Letter to Anthony Whitting, Philadelphia, December 16, 1792," in *Papers of George Washington, Presidential Series*, vol. 11, edited by Christine S. Patrick (Charlottesville: University of Virginia Press, 2002), 519.
26 Campbell, "'My Constant Companion,'" 63.
27 Campbell, "'My Constant Companion,'" 54, 63.
28 Campbell, "'My Constant Companion,'" 67.
29 Sarah Hand Meacham, "Pets, Status, and Slavery in the Late-Eighteenth-Century Chesapeake," *Journal of Southern History* 77 (August, 2011): 524.
30 Campbell, "'My Constant Companion,'" 54.
31 "Sheep and Dogs," *Charleston Courier*, March 26, 1827, 1.
32 "The City of Dogs," *Georgian* (Savannah), April 28, 1832, 2.
33 "Sheep-Killing Dogs Inquiries," *Southern Cultivator*, June 1848, 88.
34 Americus, "Dogs," *Southern Cultivator*, August 1849, 119.
35 "Hogs vs. Dogs," *Southern Planter*, October 1858, 627.
36 Wm. M. Morton, "Communicated: Mr. Editor," *Daily Constitutionalist*, May 17, 1861, 3.
37 Meacham, "Pets, Status, and Slavery in the Late-Eighteenth-Century Chesapeake," 548–549.

38 Meacham, "Pets, Status, and Slavery in the Late-Eighteenth-Century Chesapeake," 526. Meacham states that many colonists in the mid-eighteenth-century Chesapeake region aspired to emulate the British gentry. For more on Britain's growing interest in pet keeping during this time, see: Ingrid H. Tague, *Animal Companions: Pets and Social Change in Eighteenth-Century Britain* (University Park: Penn State Press, 2015).
39 John Quincy Adams, "Letter to John Adams, L'Orient, May 18, 1785," in *Adams Family Correspondence*, vol. 6, edited by Richard Alan Ryerson (Cambridge, MA: Harvard University Press, 1992), 152–153; George Washington, "Diary Entry, August 24, 1785," in *The Diaries of George Washington*, vol. 4, edited by Donald Jackson and Dorothy Twohig (Charlottesville: University of Virginia Press, 1997), 186.
40 "George Washington to Charles Carter, Mount Vernon, February 5, 1788," in *The Papers of George Washington, Confederation Series*, vol. 6, 87.
41 Meacham, "Pets, Status, and Slavery in the Late-Eighteenth-Century Chesapeake," 554; Katherine C. Grier, *Pets in America: A History* (Chapel Hill: University of North Carolina Press, 2006), 33.
42 Peter Minor, "Letter to Thomas Jefferson, Ridgway, September 17, 1811," in *The Papers of Thomas Jefferson, Retirement Series*, vol. 4, edited by J. Jefferson Looney (Princeton: Princeton University Press, 2004), 161.
43 Thomas Jefferson, "Letter to Peter Minor, Monticello, September 24, 1811," in *The Papers of Thomas Jefferson, Retirement Series*, vol. 4, edited by J. Jefferson Looney (Princeton: Princeton University Press, 2004), 170.
44 Albemarle County Residents, "Petition to Virginia General Assembly, before December 19, 1811," in *The Papers of Thomas Jefferson, Retirement Series*, vol. 4, edited by J. Jefferson Looney (Princeton: Princeton University Press, 2004), 346.
45 A Subscriber, "To The Printer," in *Royal Georgia Gazette* (Savannah), April 26, 1781.
46 "Dogs," *Times* (Charleston), July 10, 1819, 4. See also: "Sheep and Dogs," *Charleston Courier*, March 26, 1827, 1.
47 "To Guard Sheep from being Killed by Dogs," *Farmer's Register* (May 31, 1842): 205; Amer. Farmer, "To Guard Sheep from being Killed by Dogs," *Southern Agriculturalist* 2 (September 1842): 464e.
48 Homespun, "Shall Dogs be Taxed?" *Southern Cultivator*, October 1852, 316.
49 "A Tax on Female Dogs," *Southern Planter* 16 August, 1856, 248.
50 Vindicator, "For the Mercury: 'Dogs and the Bill to Prevent Dogs Running at Large,' &c," *Charleston Mercury*, August 13, 1859, 1.
51 Wm. M. Morton, "Communicated: Mr. Editor," *Daily Constitutionalist*, May 17, 1861, 3.
52 "Slaughter of Dogs," *Macon Telegraph*, June 16, 1860, 3.
53 Oscar Montgomery Lieber, *Reports of the Geognostic Survey of South Carolina*, vol. 1 (Columbia: Gibbes, 1858), 126.
54 "Importation of Blood Horses," *Alexandria Gazette*, January 27, 1836, 3.
55 Katherine C. Mooney, *Race Horse Men: How Slavery and Freedom Were Made at the Racetrack* (Cambridge: Harvard University Press, 2014); Lara Otis, "Washington's Lost Racetracks: Horse Racing from the 1760s to the 1930s,"

*Washington History* 24 (2012): 136–154; Kenneth Cohen, "Well Calculated for the Farmer: Thoroughbreds in the Early National Chesapeake, 1790–1850," *Virginia Magazine of History and Biography* 115 (2007): 382; John Eisenberg, *The Match Race: When North Met South in America's First Sports Spectacle* (Boston: Houghton Mifflin, 2006).

56 Frank L. Owsley, "The Pattern of Migration and Settlement on the Southern Frontier," *Journal of Southern History* 11 (May, 1945): 153–154.

57 Department of State, *Compendium of the Enumeration of the Inhabitants and Statistic of the United States* (Washington, DC: Thomas Allen, 1841), 359.

58 Eric Stoykovich, "The Culture of Improvement in the Early Republic: Domestic Livestock, Animal Breeding, and Philadelphia's Urban Gentlemen, 1820–1860," *Pennsylvania Magazine of History and Biography* 134 (January, 2010): 31–58; James Westfall Thompson, *A History of Livestock Raising in the United States, 1607–1860* (Washington, DC: United States Department of Agriculture, 1942): 68; Eric Carlos Stoykovich, "In the National Interest: Improving Domestic Animals and the Making of the United States, 1815–1870," PhD Dissertation, University of Virginia, 2009.

59 Verlyn Klinkenborg, "Six Million Mules," *American History* 47 (2012): 60–65; Jack Temple Kirby, *Rural Worlds Lost: The American South, 1920–1960* (Baton Rouge: Louisiana State University Press, 1986), 196.

60 Harvey Riley, *The Mule: A Treatise on the Breeding, Training, and Uses to Which He May be Put* (Philadelphia: Claxton, 1869), 36.

61 Thompson, *A History of Livestock Raising in the United States*, 68.

62 Robert Byron Lamb, *The Mule in Southern Agriculture* (Berkeley: University of California Press, 1963), 5–6.

63 Johann David Schoepf, *Travels in the Confederacy, 1783–1784*, translated and edited by Alfred J. Morrison (Philadelphia: William J. Campbell, 1911), 48.

64 Olmsted, *A Journey in the Seaboard Slave States*, 47.

65 George B. Ellenberg, "African Americans, Mules, and the Southern Mindscape, 1850–1950," *Agricultural History* 72 (spring, 1998): 381–398.

66 Lamb, *The Mule in Southern Agriculture*, 8.

67 "The Greenville S. C. Mountaineer," *Alexandria Gazette*, October 18, 1831, 3.

68 "Quantity of Livestock," *Georgian* (Savannah), February 25, 1832, 2.

69 Lamb, *The Mule in Southern Agriculture*, 23.

70 James Silk Buckingham, *The Slave States of America*, vol. 2 (London: Fisher, Son & Co., 1842), 203–204.

71 Edmund Ruffin, *Report of the Commencement and Progress of the Agricultural Survey of South Carolina*, appendix, section VIII (Columbia: A. E. Pemberton, 1843), 7.

72 Quoted in Lamb, *The Mule in Southern Agriculture*, 17.

73 Ann Norton Greene, *Horses at Work: Harnessing Power in Industrial America* (Cambridge: Harvard University Press, 2008), 97.

74 Lamb, *The Mule in Southern Agriculture*, 28.

75 A Subscriber, "On Crossing our Bred Horse with the Wild or Prairie Horse," *American Turf Register and Sporting Magazine* 4 (June, 1833): 501.

76 J. S. Skinner, "Letter to Gen. Gratiot," *American Turf Register and Sporting Magazine* 5 (January, 1834): 221. (Italics in the original.)

77 "Trial of Speed," *Spirit of the Times: a Chronicle of the Turf, Agriculture, Field Sports, Literature and the Stage* (February 18, 1843): 605. (Italics in the original.)
78 "Art. XIV – Florida," *The Port-Folio* 11 (March, 1821): 99.
79 Comte de Castelnau, "Essay on Middle Florida, 1837–1838," (translated by Arthur R. Seymour) *Florida Historical Quarterly* 26 (January 1948): 230.
80 T. Holmes, "Some Account of the Wild Horses of the Sea Islands of Virginia and Maryland," *Farmer's Register* 3 (November, 1835): 417.
81 J. H. Guenebault, "A Foreigner's First Glimpses of Georgia," *The Southern and Western Monthly Magazine and Review* 2 (August, 1845): 107.
82 "A Trip to Some Sea Islands," *The Farmers Register* 3 (January, 1836): 531.
83 "Natural History: Horse," *American Turf Register and Sporting Magazine* 1 (September 1829): 19.
84 Brooks Miles Barnes and Barry R. Truitt, (eds.), *Seashore Chronicles: Three Centuries of the Virginia Barrier Islands* (Charlottesville: University of Virginia Press, 1999): 41.
85 Holmes, "Some Account of the Wild Horses of the Sea Islands of Virginia and Maryland," 417–419.
86 Holmes, "Some Account of the Wild Horses of the Sea Islands of Virginia and Maryland," 417–419.
87 "People of Accomac," *Hartford Daily Courant*, April 29, 1848, 2.
88 Edmund Ruffin, *Agricultural, Geological, and Descriptive Sketches of Lower North Carolina, and the Similar Adjacent Lands* (Raleigh: Institute for the Deaf & Dumb & the Blind, 1861), 131–133.
89 S. M. Shepard, *The Hog in America, Past and Present* (Indianapolis: Swine Breeders' Journal, 1886), 224–225.
90 White, "From Capitalist Pigs to Globalized Pig Breeds," 104.
91 Sam White, "From Capitalist Pigs to Globalized Pig Breeds: A Study in Animal Cultures and Evolutionary History," *Environmental History* 16 (January, 2011): 94–120; Thompson, *A History of Livestock Raising in the United States*, 10.
92 Schoepf, *Travels in the Confederacy*, 246.
93 Shepard, *The Hog in America*, 257.
94 Sam B. Hilliard, "Pork in the Ante-Bellum South: The Geography of Self-Sufficiency," *Annals of the Association of American Geographers* 59 (September, 1969): 464.
95 Hilliard, "Pork in the Ante-Bellum South," 464.
96 Buckingham, *The Slave States of America*, vol. 2, 234.
97 Buckingham, *The Slave States of America*, vol. 2, 234.
98 Olmsted, *A Journey in the Seaboard Slave States*, 65–66.
99 Olmsted, *A Journey in the Seaboard Slave States*, 79.
100 Sam Hilliard, "Hog Meat and Cornpone: Food Habits in the Ante-bellum South," *Proceedings of the American Philosophical Society* 113 (February 20, 1969): 2.
101 S. Jonathan Bass, "How 'Bout a Hand for the Hog: The Enduring Nature of the Swine as a Cultural Symbol of the South," *Southern Cultures* 1 (1995): 301; Sam Hilliard, *Hog Meat and Hoe Cake: Food Supply in the Old South, 1840–1860* (Carbondale: Southern Illinois University Press, 1972); Eugene D. Genovese, *Roll, Jordan, Roll: The World the Slaves Made*

(New York: Vintage: 1976); Eugene D. Genovese, "Livestock in the Slave Economy of the Old South: A Revised View," *Agricultural History* (1962): 143–149; Ras Michael Brown, *African-Atlantic Cultures and the South Carolina Lowcountry* (Cambridge: Cambridge University Press, 2012); Royce Shingleton, "The Republic of Porkdom Revisited: A Note, Comment on the Article 'Hog Meat and Cornpone: Food Habits in the Antebellum South,'" *Proceedings of the American Philosophical Society* 114 (1970): 1–13.

102 Hillard, "Pork in the Ante-Bellum South," 462.
103 Buckingham, *The Slave States of America*, vol. 2, 234.
104 Quoted in Hilliard, "Hog Meat and Cornpone," 4.
105 Buckingham, *The Slave States of America*, vol. 2, 327.
106 Michael D. Thompson "Everything but the Squeal": Pork as Culture in Eastern North Carolina," *North Carolina Historical Review* 82 (October, 2005): 480.
107 Andrews, "Beasts of the Southern Wild," 21–47; Mart A. Stewart, "Slavery and the Origins of African American Environmentalism," in *To Love the Wind and the Rain: African Americans and Environmental History*, edited by Dianne D. Glave and Mark Stoll (Pittsburgh: University of Pittsburgh Press, 2005), 9–20; Jeff Forret, "Slaves, Poor Whites, and the Underground Economy of the Rural Carolinas," *Journal of Southern History* 70 (November, 2004): 783; Loren Schwininger, "Slave Independence and Enterprise in South Carolina, 1780–1865," *South Carolina Historical Magazine* 93 (April, 1992): 103; David W. Dangerfield, "Turning the Earth: Free Black Yeomanry in the Antebellum South Carolina Lowcountry," *Agricultural History* 89 (spring, 2015): 210; Karl Jacoby, "Slaves by Nature? Domestic Animals and Human Slaves," *Slavery & Abolition* 15 (1994): 89–99; Philip D. Morgan, "The Ownership of Property by Slaves in the Mid-Nineteenth-Century Low Country," *Journal of Southern History* 49 (1983): 399–420; Philip D. Morgan, "Work and Culture: The Task and the World of Lowcountry Blacks, 1700 to 1880," *William and Mary Quarterly* 39 (1982): 563–599.
108 Olmsted, *A Journey in the Seaboard Slave States*, 422, 439. For more on the topic of slaves who owned property, see: Dylan Penningroth, "Slavery, Freedom, and Social Claims to Property among African Americans in Liberty County, Georgia, 1850–1880," *Journal of American History* 84 (September, 1997): 405–435.
109 Schweninger, "Slave Independence and Enterprise in South Carolina," 101; Joyce E. Chaplin, *An Anxious Pursuit: Agricultural Innovation and Modernity in the Lower South, 1730–1815* (Chapel Hill: University of North Carolina Press, 1996), 325.
110 Quoted in John Hope Franklin, "Slaves Virtually Free in Ante-Bellum North Carolina," *Journal of Negro History* 28 (July, 1943): 308–309.
111 Forret, "Slaves, Poor Whites, and the Underground Economy of the Rural Carolinas," 783–824. See also: Ras Michael Brown, *African-Atlantic Cultures and the South Carolina Lowcountry* (Cambridge University Press, 2012); 156; Scott Giltner, "Slave Hunting and Fishing in the Antebellum South," in *To Love the Wind and the Rain: African Americans and Environmental History*, edited by Dianne D. Glave and Mark Stoll (Pittsburgh: University of Pittsburgh

Press, 2005), 21–36; Penningroth, "Slavery, Freedom, and Social Claims to Property among African Americans in Liberty County, Georgia, 1850–1880," 412–413; Philip D. Morgan, "The Ownership of Property by Slaves in the Mid-Nineteenth-Century Low Country," *Journal of Southern History* 49 (August, 1983): 399–420.

112 Ryan Quintana, "Planners, Planters, and Slaves: Producing the State in Early National South Carolina," *Journal of Southern History* 81 (February, 2015): 79–116; Peter McCandless, *Slavery, Disease, and Suffering in the Southern Lowcountry* (New York: Cambridge University Press, 2011); Karen B. Bell, "Rice, Resistance, and Forced Transatlantic Communities: (Re)Envisioning the African Diaspora in Lowcountry Georgia, 1750–1800," *Journal of African American History* 95 (2010): 174; Laura F. Edwards, "Southern History as U. S. History," *Journal of Southern History* 75 (August, 2009): 556; Kenneth Stamp, *The Peculiar Institution: Slavery in the Antebellum South* (Berkeley: University of California, 1956).

113 Jack Temple Kirby, *Poquosin: A Study of Rural Landscape and Society* (Chapel Hill: University of North Carolina Press, 1995), 147.

114 J. F. D. Smyth, *A Tour in the United States of America*, vol. 1 (Dublin: Price, Moncrieffe, 1784), 101–102. See also: Marvin L. Michael Kay and Lorin Lee Cary, "Slave Runaways in Colonial North Carolina, 1748–1775," *North Carolina Historical Review* 63 (1986): 5.

115 Schweninger, "Slave Independence and Enterprise in South Carolina, 1780–1865," 117. See also: Tim Lockley and David Dodding, "Maroon and Slave Communities in South Carolina before 1865," *South Carolina Historical Magazine* 113 (2012): 131; Timothy James Lockley, *Maroon Communities in South Carolina: A Documentary Record* (Columbia: University of South Carolina Press, 2009); David Cecelski, *The Waterman's Song: Slavery and Freedom in Maritime North Carolina* (Chapel Hill: University of North Carolina Press, 2001), 131; John Hope Franklin and Loren Schweninger, *Runaway Slaves: Rebels on the Plantation* (New York: Oxford University Press, 2000): 88; Kirby, *Poquosin*, 168–172.

116 Drew Addison Swanson, "Fighting over Fencing: Agricultural Reform and Antebellum Efforts to Close the Virginia Open Range," *Virginia Magazine of History and Biography* 117 (2010): 129. Notably, Thomas Jefferson, who owned hundreds of slaves, opposed the enclosure movement. Mark Sturges, "Enclosing the Commons: Thomas Jefferson, Agrarian Independence, and Early American Land Policy, 1774–1789," *Virginia Magazine of History and Biography* 119 (2011): 42–74.

117 John Taylor, *Arator*, second edition (Georgetown: J.M. Carter, 1814), 173, 183, 192, 197; Kirby, *Mockingbird Song*, 80.

118 Swanson, "Fighting over Fencing," 108–109.

119 Philander, "Enormous Losses caused by the Fence Law in Virginia," *Farmer's Register* 1 (March, 1834): 633.

120 Jeremiah, "Comparative View of the Rights of Citizens, and of Hogs, under the Fence Law of Virginia," *Farmer's Register* 2 (December, 1834): 455.

121 Jeremiah, "Stump and Barrel Legislation," *Farmer's Register* 3 (June, 1835): 126.

122 Kirby, *Rural Worlds Lost*, 116.

123 Virginia Lower, "The Fence Law," *Southern Planter* 7 (May, 1847): 134.
124 Virginia Lower, "The Fence Law," 134.
125 "The Chief Tax in Virginia," *Southern Planter* 7 (November, 1847): 344.
126 "Extracts from Private Correspondence," *Farmer's Register* 2 (September, 1834): 255.
127 Edmund Ruffin, Jr., William M. Tate, and Richard Irby, "Report to the Farmers' Assembly on the Law of Enclosures," *Southern Planter* 18, 3 (March 1858): 170.
128 Swanson, "Fighting over Fencing," 128.
129 Robert E. Gallman, "Pork Production and Nutrition during the Late Nineteenth Century: A Weighty Issue Visited Yet Again," *Agricultural History* 69 (1995): 592–606; Timothy Cuff, "A Weighty Issue Revisited: New Evidence on Commercial Swine Weights and Pork Production in Mid-Nineteenth Century America," *Agricultural History* 66 (1992): 55–74; Sam B. Hilliard, "Pork in the Ante-Bellum South: The Geography of Self-Sufficiency," *Annals of the Association of American Geographers* 59 (September, 1969): 465–466; Eugene D. Genovese, "Livestock in the Slave Economy of the Old South: a Revised View," *Agricultural History* 36 (July, 1962): 147–148.
130 U. S. Department of Commerce, *Historical Statistics of the United States: Colonial Times to 1790*, part 1 (Washington, DC: United States Government Printing Office, 1976), 24–37. Note that this number does not include the relatively small number of people living in Florida. Georgia's tally also includes the land that would later become Alabama.
131 Their numbers would have been even higher, but the demand for slave labor changed in dramatic fashion after Congress outlawed the importation of slaves from Africa in 1807. Most significantly, the law created an internal slave trade, as planters in Alabama, Mississippi, and Louisiana imported hundreds of thousands of slaves from Virginia and South Carolina to work on the growing number of cotton and sugar plantations in the "Old Southwest." David Eltis, "The U.S. Transatlantic Slave Trade, 1644–1867," *Civil War History* 54 (December, 2008): 347–378; Steven Deyle, *Carry Me Back: The Domestic Slave Trade in American Life* (New York: Oxford University Press, 2005), 4, 17; David L. Lightner, *Slavery and the Commerce Power: How the Struggle against the Interstate Slave Trade Led to the Civil War* (New Haven: Yale University Press, 2006); Damian Alan Pargas, *Slavery and Forced Migration in the Antebellum South* (New York: Cambridge University Press, 2014), 22; Deyle, *Carry Me Back*, 22; Adam Rothman, *Slave Country: American Expansion and the Origins of the Deep South* (Cambridge: Harvard University Press, 2005). Meanwhile, somewhere between one third and one half of the European Americans living between Virginia and Florida belonged to slave-owning households in 1850. Department of Commerce and Labor, *A Century of Population Growth: From the First Census of the United States to the Twelfth, 1790–1900* (Washington, DC: Government Printing Office, 1909), 138–139.
132 Hilliard, "Pork in the Ante-Bellum South," 466–467. This ration was by no means universal. For example, Charles Ball almost never ate meat when he was enslaved in South Carolina. Andrews, "Beasts of the Southern Wild," 26.

133 Hilliard, "Pork in the Ante-Bellum South," 461–480; Michael D. Thompson, "Hog Production," *New Encyclopedia of Southern Culture*, vol. 11: *Agriculture and Industry*, edited by Melissa Walker and James C. Cobb (Chapel Hill: University of North Carolina Press, 2008), 167.

134 Hilliard, "Pork in the Antebellum South," 461, 466–467; John Campbell, "Southwestern Road," *Richmond Whig*, December 18, 1840, 2.

135 Elizabeth L. Parr, "Kentucky's Overland Trade with the Antebellum South," *Filson Club History Quarterly* 2 (October, 1927–July, 1928), 73.

136 "Quantity of Livestock," *Georgian* (Savannah), February 25, 1832, 2; Parr, "Kentucky's Overland Trade with the Antebellum South," 77; Charles T. Leavitt, "Transportation and the Livestock Industry of the Middle West to 1860," *Agricultural History* 8 (January, 1934): 29; Genovese, "Livestock in the Slave Economy of the Old South: a Revised View," 143–144; Wilma A. Dunaway, *The First American Frontier: Transition to Capitalism in Southern Appalachia, 1700–1860* (Chapel Hill: University of North Carolina Press, 1996), 206.

137 Dunaway, *The First American Frontier*, 218–219.

138 Edmund Cody Burnett, "Hog Raising and Hog Driving in the Region of the French Broad," *Agricultural History* 20 (April, 1946): 86–103.

139 "A Winter in the South," *Harper's New Monthly Magazine*, vol. XV, no. LXXXIX (October, 1857): 594–595; Hilliard, "Pork in the Antebellum South," 474.

140 R. Ben Brown, "Free Men and Free Pigs: Closing the Southern Range and the American Property Tradition," *Radical History Review* 108 (2010): 118; Bass, "'How 'bout a Hand for the Hog," 304; John Solomon Otto, *Southern Agriculture during the Civil War Era, 1860–1880* (Westport: Greenwood Press, 1994): 17; Forrest McDonald and Grady McWhiney, "The Antebellum Southern Herdsmen: a Reinterpretation," *Journal of Southern History* 41 (May, 1975): 147.

141 *Compendium of the Tenth Census* (Washington, DC: Government Printing Office, 1883), 681.

142 Willard Range, *A Century of Georgia Agriculture, 1850–1950* (Athens: University of Georgia Press, 1954), 18.

143 "Thoughts for the Crisis," *Southern Cultivator*, 19, 1 (January, 1861): 10. See also: Andrews, "Beasts of the Southern Wild," 25.

144 "Plantation Economy, &c," *Southern Cultivator*, 19, 3 (March 1861): 73.

145 G. D. Harmon, "Letter 1," *Southern Cultivator*, 19, 4 (April, 1861): 114.

146 G. D. Harmon, "Supplies Stopped – Let us Make our Own," *Southern Cultivator* 19 (July, 1861): 218.

147 Michael D. Thompson, "High on the Hog: Swine as Culture and Commodity in Eastern North Carolina," PhD Dissertation, Miami University, 2000, 151.

148 Charles Francis Adams, Jr., "Letter from Charles Francis Adams, Jr. to Charles Francis Adams, Sr., November 30, 1862," in *A Cycle of Adams Letters, 1861–1865*, edited by Worthington Chauncey Ford (Boston: Houghton, Mifflin & Co., 1920), 201.

149 George Henry Gordon, "Memoir of George Henry Gordon," in *A War Diary of Events in the War of the Great Rebellion, 1863–1865* (Boston: James R. Osgood & Company, 1882), 57.

150 Henry Harrison Eby, "Memoir of Henry Harrison Eby," in *Observations of an Illinois Boy in Battle, Camp and Prisons, 1861 to 1865* (Mendota: Eby, 1910), 134.
151 Quoted in Paul Wallace Gates, *Agriculture and the Civil War* (New York: Knopf, 1965), 7.
152 Richard Lee Fulgram, *The Hogs of Cold Harbor: the Civil War Saga of Private Johnny Hess, CSA* (Pittsburgh: Whitmore, 2005).
153 "Hog Raising," *Southern Cultivator*, 20, 11–12 (November, 1862): 210.
154 "Answers to Correspondents," *Southern Cultivator*, 21, 11–12 (November/December, 1863): 129.
155 Mark Fiege, *Republic of Nature* (Seattle: University of Washington Press, 2012), 204.
156 John E. Clarke, Jr. *Railroads in the Civil War: The Impact of Management on Victory and Defeat* (Baton Rouge: Louisiana State Press, 2001); Charles W. Ramsdell, "The Confederate Government and the Railroads," *American Historical Review* 22 (July, 1917): 798.
157 Leavitt, "Transportation and the Livestock Industry of the Middle West to 1860," 26–27.
158 E. Z. Russell, et al., "Hog Production and Marketing," *Yearbook of the Department of Agriculture, 1922* (Washington, DC: Government Printing Office, 1923), 189.
159 William Cronon, *Nature's Metropolis: Chicago and the Great West* (New York: Norton and Co., 1992), 230.
160 Ted Steinberg, *Down to Earth: Nature's Role in American History* (Oxford: Oxford University Press, 2002), 95.
161 Colonel Samuel Ringwalt, "The Horse – from Practical Experience in the Army," in *Report of the Commission of Agriculture for the Year 1866* (Washington, DC: Government Printing Office, 1867), 324.
162 Steinberg, *Down to Earth*, 90.
163 *Compendium of the Tenth Census*, 675–676.
164 Charles W. Ramsdell, "General Robert E. Lee's Horse Supply, 1862–1865," *American Historical Review* 35 (July, 1930): 759.
165 John Daniel Imboden, "Memoir of John Daniel Imboden," in *The Century War Series, Vol. 2: Battles and Leaders of the Civil War* (New York: Century Co., 1887).
166 J. J. McDaniel, "Diary of J. J. McDaniel, September, 1862," in *Diary of Battles, Marches and Incidents of the Seventh S.C. Regiment* (Privately published, 1862).
167 Ramsdell, "General Robert E. Lee's Horse Supply," 759.
168 Spencer Jones, "The Influence of Horse Supply upon Field Artillery in the American Civil War," *Journal of Military History* 74 (2010): 371.
169 Jones, "The Influence of Horse Supply upon Field Artillery in the American Civil War," 369.
170 Ramsdell, "General Robert E. Lee's Horse Supply," 760.
171 Fiege, *Republic of Nature*, 208.
172 Robert E. Lee, "Letter to William N. Pendleton, April 25, 1863," in *The War of the Rebellion: a Compilation of the Official Records of the Union and Confederate*

*Armies*, series I, vol. xxv, part II (Washington, DC: Government Printing Office, 1889), 749.
173 This estimate includes soldiers who died on the battlefield, those who died from diseases that they contracted in camp, and those who fell victim to guerrilla warfare. J. David Hacker, "A Census-Based Count of the Civil War Dead," *Civil War History* 57 (December 2011): 307–348.
174 *Compendium of the Tenth Census*, 675–676, 681; Isaac Newton, "Report of the Commissioner of Agriculture," in *Report of the Commission of Agriculture for the Year 1866* (Washington, DC: Government Printing Office, 1867), 13.
175 Drew Gilpin Faust, "Equine Relics of the Civil War," in *Southern Cultures: The Fifteenth Anniversary Reader*, edited by Harry L. Watson and Larry J. Griffin (Chapel Hill: University of North Carolina Press, 2008), 391–408.

## 5 Anthropogenic Improvement and Assaults on Ferality

1 *USDA Census of Agriculture, 1945*, vol. 2, part 7 (Washington, DC: Government Printing Office, 1947), 390.
2 Samuel Mills Tracy, *Hog Raising in the South* (Washington, DC: U. S. Department of Agriculture, 1899), 5.
3 S. M. Shepard, *The Hog in America, Past and Present* (Indianapolis: Swine Breeders' Journal, 1886), 19, 225.
4 E. Z. Russell, et al., "Hog Production and Marketing," *Yearbook of the Department of Agriculture, 1922* (Washington, DC: Government Printing Office, 1923), 182, 209, 255.
5 Quoted in: William Cronon, *Nature's Metropolis: Chicago and the Great West* (New York: Norton and Company, 1991), 226. For more on the rise of the meat industry in the Midwest, see: Kristin Hoganson, "Meat in the Middle: Converging Borderlands in the U. S. Midwest, 1865–1900," *Journal of American History* (March, 2012): 1025–1051; Margaret Walsh, *The Rise of the Midwestern Meat Packing Industry* (Lexington: University of Kentucky Press, 1982); Margaret Walsh, "From Pork Merchant to Meat Packer: The Midwestern Meat Industry in the Mid-Nineteenth Century," *Agricultural History* 56 (January, 1982): 127–137; Margaret Walsh, "Pork Packing as a Leading Edge of Midwestern Industry, 1835–1875," *Agricultural History* 51 (October, 1977): 702–717.
6 James Westfall Thompson, *A History of Livestock Raising in the United States, 1607–1860* (Washington, DC: United States Department of Agriculture, 1942), 10; Shepard, *The Hog in America*, 15, 19.
7 Cronon, *Nature's Metropolis*, 230; Bass, "How 'Bout a Hand for the Hog," *Southern Cultures* 1 (1995): 310; Roger Horowitz, *Putting Meat on the American Table: Taste, Technology, Transformation* (Baltimore: Johns Hopkins University Press, 2006), 43–74.
8 John Majewski and Viken Tchakerian, "The Environmental Origins of Shifting Cultivation: Climates, Soils, and Disease in the Nineteenth-Century US South," *Agricultural History* 81 (fall, 2007): 536.
9 Amrys Williams, "Cultivating Modern America: 4-H Clubs and Rural Development in the Twentieth Century," PhD Dissertation, University of

Wisconsin-Madison, 2012. See also: "Modern Farm Methods as Applied in the South," *The New Enterprise* (Madison, FL), June 24, 1909, 6. Regarding the South's questionable hog-and-hominy reputation, see: *Atlanta Constitution*, September 21, 1926, 6.

10 Perry Van Ewing, *Southern Pork Production* (New York: Orange Judd Company, 1918), 8–9.
11 Van Ewing, *Southern Pork Production*, 30.
12 Drew Addison Swanson, "Fighting over Fencing: Agricultural Reform and Antebellum Efforts to Close the Virginia Open Range," *Virginia Magazine of History and Biography* 117 (2010): 128–129.
13 "Chap. 14 – An Act to repeal the Fence Law of Virginia as to certain Counties, and to authorize the County Courts to dispense with Enclosures in other Counties, October 3, 1862," *Acts of the General Assembly of the Commonwealth of Virginia* (Richmond: William F. Ritchie, 1862), 20–21.
14 "The Fence Law," *Daily Constitutionalist* (Augusta, Georgia), January 31, 1869, 2; "The Fence Law Again," *Daily Constitutionalist* (Augusta, Georgia), February 13, 1869, 2; "The Fence Law," *Augusta Chronicle*, June 25, 1878, 1.
15 *The Fence Question in Southern States as Related to General Husbandry and Sheep Raising* (Worcester, Mass: Washburn & Moen Manufacturing Co. 1881), 11.
16 Brook Blevins, "Fence/Stock Laws," in *New Encyclopedia of Southern Culture, volume 11: Agriculture and Industry*, edited by Melissa Walker and James C. Cobb (Chapel Hill: University of North Carolina Press, 2008): 159; R. Ben Brown, "Free Men and Free Pigs: Closing the Southern Range and the American Property Tradition," *Radical History Review* 108 (2010): 117–137; J. Crawford King, Jr., "The Closing of the Southern Range: An Exploratory Study," *Journal of Southern History* 48 (February, 1982): 53–70.
17 Brown, "Free Men, Free Pigs," 121, 124.
18 "A No-fence Law in this County," *The News and Courier* (Charleston), December 8, 1879, 1. That same year, wire fencing was first introduced to the South, obviating some of the arguments against timber supply. King, "The Closing of the Southern Range," 57; Lyn Ellen Bennett and Scott Abbott, "Barbed and Dangerous: Constructing the Meaning of Barbed Wire in Late Nineteenth-Century America," *Agricultural History* 88 (fall, 2014): 566–590.
19 General William M. Browne, "Agriculture Department," *Macon Telegraph*, January 27, 1880, 3. "The Fence Law," *Augusta Chronicle*, January 14, 1882, 1; R. Ben Brown, "The Southern Range: A Study in Nineteenth-Century Law and Society," PhD dissertation, University of Michigan, 1993, 293–294; Brown, "Free Men, Free Pigs," 132; Steven Hahn, *Roots of Southern Populism: Yeoman Farmer and the Transformation of the Georgia Upcountry, 1850–1890* (Oxford: Oxford University Press, 1983); Shawn Everett Kantor, *Politics and Property Rights: The Closing of the Open Range in the Postbellum South* (Chicago: University of Chicago Press, 1998).
20 Robert R. Dykstra, "Town–Country Conflict: A Hidden Dimension in American Social History," *Agricultural History* 38 (October, 1964): 195–204.
21 "The Fence Law," *Augusta Chronicle*, June 25, 1878, 1; Brown, "Free Men and Free Pigs," 127; "The Moore County Trouble," *Charlotte Observer*, August 17, 1897, 4.

22 B. F. Keith, "Advocates No-Fence Law," *Charlotte Daily Observer*, April 19, 1908, 4; "The No-Fence Fanatics," *Tampa Tribune*, August 28, 1895, 2.
23 Ted Steinberg, *Down to Earth: Nature's Role in American History* (Oxford: Oxford University Press, 2002), 12; "The No-Fence Law," *Winston-Salem Journal*, September 15, 1908, 4; "The Razorback Hog," *Charlotte Observer*, April 15, 1910, 11; "More Meat-Makers, Fewer Razor-backs," *The Pickens Sentinel* (Pickens, South Carolina), March 2, 1916, 7; "Hogs Disappear," *Richmond Times-Dispatch*, May 7, 1905, 12; "The Razorback Hog," *Forest and Stream*, March 19, 1910, 2.
24 "Where, and Oh! Where are the Razorback Hogs?" *Meat and Livestock Digest* (September, 1922): 2. National Archives (College Park) / Record Group 16 – Records of the Offices of the Secretary of Agriculture / General Correspondence of the Office of the Secretary, 1906–1970 / Animals – Appropriations 1922 / Box no. 876 / Folder: Animals-Hogs.
25 W. F. Yocum, *Florida Agricultural Experiment Station, Report for Financial Year ending June 30, 1901* (DeLand, FL: Painter & Company, 1901), *Forest and Stream Frank Leslie's Popular Monthly Forest and Stream* 22; S. C. C., "Swine in Florida," *Forest and Stream*, November 23, 1876, 246; "A Hunting Party in Florida," *Frank Leslie's Popular Monthly* (February 1881), 156; "The Sportsman Tourist," *Forest and Stream* (November 24, 1881), 325.
26 "A Strange Change," *The Atlanta Constitution*, "August 14, 1887, 13; Charles Sloan Reid, "Swine Department: The Mast Hog," *National Stockman and Farmer 19* (December 5, 1895 ), 4."
27 "Where, and Oh! Where are the Razorback Hogs?" *Meat and Livestock Digest* (September 1992): 2. National Archives (College Park) / Record Group 16 – Records of the Offices of the Secretary of Agriculture / General Correspondence of the Office of the Secretary, 1906–1970 / Animals – Appropriations 1922 / Box no. 876 / Folder: Animals-Hogs; Tobe Hodge, "Razor-Backs," *The American Magazine*, vii, 2 (December, 1887): 254; "Virginia's Wild Hogs: One of the Most Wonderful Animals Yet Evoluted," *St. Louis Post-Dispatch*, January 21, 1888, 6; J. M. Murphy, "Florida Razorbacks," *Outing* 19 (November 1891): 117; Shepard, *The Hog in America*, 20; Reid, "Swine Department: The Mast Hog," 4; Leonidas Hubbard, "Hunting Wild Hogs as a Sport," *New York Times*, September 1, 1901, sm9.
28 John F. Reiger, *American Sportsmen and the Origins of Conservation* (Corvallis: Oregon State University Press, 2001); Jack Temple Kirby, *Mockingbird Song: Ecological Landscapes of the South* (Chapel Hill: University of North Carolina Press, 2006), 164; Scott Giltner, *Hunting and Fishing in the New South: Black Labor and White Leisure after the Civil War* (Baltimore: Johns Hopkins University Press, 2008), 118; Walter F. Mickle, "Ft. Myers, Fla.," *Forest and Stream* 37 (November 12, 1891): 327. On the rise of other outdoor sports in the American South during the Progressive era, see: Kevin Kokomoor, "The 'Most Strenuous of Anglers' Sports is Tarpon Fishing': The Silver King as Progressive Era Outdoor Sport," *Journal of Sport History* 37 (2010): 347–364.
29 *Hunting and Fishing on the Sea Board Air Line Railway* (N.P.: Seaboard Air Line Rail Way, c. 1900).

30 Steinberg, *Down to Earth*, 107; *Hunting and Fishing on the Seaboard Air Line Railway* (N.P.: Seaboard Air Line Railway, c. 1900), 1, 3 (Copy located at Harvard University Library); "Industrial Conditions Improve Big Game Hunting," *Omaha Daily Bee* (Omaha, Nebraska), November 9, 1902; "Hunting Days in Florida," *New York Times*, May 30, 1897, 17; Minnie Moore-Wilson, "Hunting in a Florida Jungle," *Forest and Stream* LIX (October 25, 1902), 324.

31 Hubbard, "Hunting Wild Hogs as a Sport," sm9; Reid, "Swine Department: The Mast Hog," 4; "Hunting Wild Hogs," *Dona Ana County Republican* (Las Cruces, NM), December 21, 1901, 2.

32 James H. Tuten, *Lowcountry Time and Tide: The Fall of the South Carolina Rice Kingdom* (Columbia: University of South Carolina Press, 2010); Gilbert C. Fite, "Southern Agriculture since the Civil War: An Overview," *Agricultural History* 53 (January, 1979): 8; Scott Giltner, *Hunting and Fishing in the New South*, 110–111. 117; Hubbard, "Hunting Wild Hogs as a Sport," sm9.

33 William Whitson, "A South Carolina Hunt," *Forest and Stream* LIII (September 30, 1899): 264; Giltner, *Hunting and Fishing in the New South*, 133.

34 Frank G. Carpenter, "The Isle of Millionaires," *Los Angeles Times*, December 27, 1896, 23; William Barton McCash and June Hall McCash, *The Jekyll Island Club: Southern Haven for America's Millionaires* (Athens: University of Georgia, 1989), 20.

35 Montgomery M. Folsom, "Game on Jekyl," *The Atlanta Constitution*; February 17, 1889, 3; "The Jekyl Island Club," *New York Times*, September 9, 1889, 5.

36 Hubbard, "Hunting Wild Hogs as a Sport," sm9; "Industrial Conditions Improve Big Game Hunting," *Omaha Daily Bee*, November 9, 1902; "The Most Beautiful and Greatest Sea Island in Florida," *Tampa Tribune*, March 22, 1925, 50; "Wild Hog Hunting found profitable in Florida Swamp," *Tampa Tribune*, May 30, 1926, 20; Giltner, *Hunting and Fishing in the New South*, 7.

37 McCash and McCash, *The Jekyll Island Club*, 21; John J. Mayer, "Taxonomy and History of Wild Pigs in the United States," in *Wild Pigs: Biology, Damage, Control Techniques and Management*, edited by John J. Mayer and Lehr Brisbin (Aiken, South Carolina: Savannah River Site: 2009); William H. Stiver and E. Kim Delozier, "Great Smoky Mountains National Park Wild Hog Control Program," in *Wild Pigs: Biology, Damage, Control Techniques and Management* edited by John J. Mayer and I. Lehr Brisbin (Aiken, South Carolina: Savannah River Site: 2009), 341–342; Scott Hart, "The Federal Diary Registered in U. S. Patent Office: Bessie the Boar, *The Washington Post*; January 29, 1938, x24; LeRoy C. Stegeman, "The European Wild Boar in the Cherokee National Forest, Tennessee," *Journal of Mammalogy* 19 (August, 1938): 280–281.

38 Ann Norton Greene, "War Horses: Equine Technology in the Civil War," in *Industrializing Organisms: Introducing Evolutionary History*, edited by Susan R. Schrepfer and Philip Scranton (New York: Routledge: 2004), 145.

39 "Wild Horses by Thousands," *Sioux City Journal*, June 1, 1900, 10.

40 Howard Pyle, "Chincoteague: the Island of Ponies," *Scribner's Monthly*, xiii, 6 (April, 1877): 737; "A Future for the Lighter," *New York Times*, October 12, 1884, 4.

Notes to pages 93–97

41 "Hundreds of Wild Ponies," *Washington Post*, February 16, 1902, 19; Pyle, "Chincoteague: the Island of Ponies," 737; "Out of the World on the Atlantic Coast," *New York Herald*, October 18, 1891.
42 "Natural History: North Carolina Wild Horses," *Forest and Stream*, October 17, 1903, 297; H. W. G., "Scenes in Carolina," *Atlanta Constitution*, August 6, 1882, 1; Melville Chater, "Motor-Coaching through North Carolina," *National Geographic*, vol. xlix, no. 5 (May, 1926): 479; "Natural History: North Carolina Wild Horses," *Forest and Stream*, October 17, 1903, 297.
43 Fred A. Olds, "The Wild Horse of the Banks," *Forest and Stream*, November 15, 1902, 384; "Wild horses of North Carolina," *Springfield Republican* (MA), December 7, 1902, 8; "The Wild Horse in America," *Forest and Stream*, Jan 3, 1903, 1.
44 "Jekyl Island," *Forest and Stream*, October 11, 1888, p. 221.
45 John Van Wormer, "An Island on the Georgia Coast," *The Cosmopolitan* 24 (January, 1898): 289; "Jekyl Island," *Forest and Stream*, October 11, 1888, 221; Montgomery M. Folsom, "Game on Jekyl," *Atlanta Constitution*, February 17, 1889, 3; Frederick A. Ober, "Dungeness, General Greene's Sea-Island Plantation," *Lippincott's Magazine of Popular Literature and Science* 26 (August, 1880): 243.
46 "Motor Vehicle Registrations, by States, 1900–1995," Federal Highway Administration, U. S. Department of Transportation. URL: www.fhwa.dot .gov/.
47 Michael L. Berger, *The Devil Wagon in God's Country: The Automobile and Social Change in Rural America, 1893–1929* (Hamden: Archon, 1979), 13. Regarding the arrival of automobiles in Mississippi, see: Corey T. Lessig, *Automobility: Social Changes in the American South, 1909–1939* (New York: Routledge, 2001). Regarding the arrival of automobiles in the rural United States, see: Ronald Kline and Trevor Pinch, "Users as Agents of Technological Change: The Social Construction of the Automobile in the Rural United States," *Technology and Culture* 37 (1996): 763–795.
48 Steve Gurr, "Toy, Tool, and Token: Views of Early Automobiling in Georgia," *Georgia Historical Quarterly* 77 (1993): 384.
49 "Automobiles and Bicycles," *Evening Post* (Charleston, SC), July 21, 1899, 8.
50 "Future of the Horseless Carriage," *Columbus Daily Enquirer* (GA), May 16, 1899, 4.
51 "Motor Vehicle Registrations, by States, 1900–1995," Federal Highway Administration, U. S. Department of Transportation. URL: www.fhwa.dot.gov/.
52 "Street Cars vs. Automobiles," *Macon Telegraph*, June 20, 1902, 2.
53 "Quarter Million Dollars in Augusta Automobiles," *Augusta Chronicle*, January 26, 1908, 10.
54 Randall L. Hall, "Before NASCAR: The Corporate and Civic Promotion of Automobile Racing in the American South, 1903–1927," *Journal of Southern History* 68 (August, 2002): 637; Alice Strickland, "Florida's Golden Age of Racing," *Florida Historical Quarterly* 45 (1967): 253–269.
55 Hall, "Before NASCAR," 638.
56 Quoted in: Hall, "Before NASCAR," 638.

57 Hall, "Before NASCAR," 642; Howard L. Preston, "The Automobile Business in Atlanta, 1909–1920: A Symbol of 'New South' Prosperity," *Georgia Historical Quarterly* 58 (1974): 262–277.
58 Hall, "Before NASCAR," 647. See also: Blaine A. Brownell, "A Symbol of Modernity: Attitudes Toward the Automobile in Southern Cities in the 1920s," *American Quarterly* 24 (March, 1972): 20–44.
59 "Automobiles in the Country," *Augusta Chronicle*, January 8, 1905, 4.
60 "Over South Carolina the Auto has Friends," *The State* (Columbia SC), October 29, 1911, part IV.
61 Berger, *The Devil Wagon in God's Country*, 42–43. Note that many farmers were already familiar with gasoline engines, which helped facilitate their adoption of gas-powered automobile. Carrie A. Meyer, "The Farm Debut of the Gasoline Engine," *Agricultural History* 87 (2013): 287–313.
62 O. E. Baker, *A Graphic Summary of Farm Machinery, Facilities, Roads, and Expenditures*, United States Department of Agriculture, Miscellaneous Publication No. 264 (Washington, DC: USDA, 1934), 4. National Archives (College Park) / Record Unit 17 – Bureau of Animal Industry / Box 5 – Records Pertaining to Horse and Dogs / Folder – Machines.
63 "Carriages for All," *The Post* (Charleston SC), June 2, 1897, 7.
64 "Progress in Automobiles," *The State* (Columbia SC), August 14, 1899, 3.
65 "Street Cars vs. Automobiles," *Macon Telegraph*, June 30, 1902, 2.
66 "Bitter against the Automobiles," *The Columbus Enquirer-Sun*, September 6, 1903, 12.
67 "Automobiles in the Country," *The Augusta Chronicle*, January 8, 1905, 4.
68 "Horses Must Learn," *Augusta Chronicle*, June 25, 1899, 8.
69 "Automobiles in the Country," *The Augusta Chronicle*, January 8, 1905, 4.
70 "The Price of Horses," *Macon Daily Telegraph*, July 23, 1907, 5.
71 *United States Census of Agriculture: 1945*, vol. 2, ch. 7 (Washington, DC: Government Printing Office, 1947), 366.
72 "Will We Continue to Need Draft Horses?" National Archives (College Park) / Record Unit 17 – Bureau of Animal Industry / Box 1 – Records Pertaining to Horse and Dogs / Folder – Horses-Breeds.
73 Ellsworth Huntington, "The Distribution of Domestic Animals," *Economic Geography* 1 (July, 1925): 145–146.
74 Paul G. Irwin, "Overview: The State of Animals in 2001," in *The State of the Animals, 2001*, edited by Deborah J. Salem and Andrew N. Rowan (Washington, DC: Humane Society Press, 2001), 8. On the mechanization of American farms during the 1920s and 1930s, see: Alan L. Olmsted and Paul W. Rhode, "The Agricultural Mechanization Controversy of the Interwar Years," *Agricultural History* 68 (1994): 35–53.
75 *Compendium of the Tenth Census* (Washington, DC: Government Printing Office, 1883), 681; *Horses, Mules, and Motor Vehicles*, United States Department of Agriculture, Statistical Bulletin No. 5 (Washington, DC: Government Printing Office, 1925), 3; *1945 Agriculture Census*, vol. 2, part 7 (Washington, DC: Government Printing Office, 1947), 366.
76 *United States Census of Agriculture: 1945*, vol. 2, ch. 7 (Washington, DC: Government Printing Office, 1947), 366.

77 "Motor Vehicle Registrations, by States, 1900–1995," Federal Highway Administration, U. S. Department of Transportation. URL: www.fhwa.dot.gov/.
78 George B. Ellenberg, *Mule South to Tractor South: Mules, Machines, and the Transformation of the Cotton South* (Tuscaloosa: University of Alabama Press, 2007); Larry Sawers, "The Mule, the South, and Economic Progress," *Social Science History* 28 (winter, 2004): 667–690; 667; Watson C. Arnold, "The Mule: The Worker that 'Can't Get No Respect,'" *Southwestern Historical Quarterly* 112 (2008): 35–50; Alan L. Olmsted and Paul W. Rhode, "Reshaping the Landscape: The Impact and Diffusion of the Tractor in American Agriculture, 1910–1960," *Journal of Economic History* 61 (2001): 663–698.
79 *United States Census of Agriculture: 1945*, vol. 2, ch. 7 (Washington, DC: Government Printing Office, 1947), 368.
80 O. E. Baker, *A Graphic Summary of Farm Machinery, Facilities, Roads, and Expenditures*, United States Department of Agriculture, Miscellaneous Publication No. 264 (Washington, DC: USDA, 1934), 4. National Archives (College Park) / Record Unit 17 – Bureau of Animal Industry / Box 5 – Records Pertaining to Horse and Dogs / Folder – Machines. Regarding the diffusion of tractor technology in the Midwest, see: Dinah Duffy Martini and Eugene Silberberg, "The Diffusion of Tractor Technology," *Journal of Economic History* 66 (2006): 354–389.
81 *United States Census of Agriculture: 1945*, vol. 2, ch. 7 (Washington, DC: Government Printing Office, 1947), 371.
82 Clay McShane and Joel A. Tarr, *The Horse in the City: Living Machines in the Nineteenth Century* (Baltimore: Johns Hopkins University Press, 2007), 14; Olmsted and Rhode, "Reshaping the Landscape: The Impact and Diffusion of the Tractor in American Agriculture, 1910–1960," 667.
83 Pete Daniels, *Standing at the Crossroads: Southern Life in the Twentieth Century* (Baltimore: Johns Hopkins University Press, 1996), 83.
84 Jack Temple Kirby, *Rural Worlds Lost: The American South, 1920–1960* (Baton Rouge: Louisiana State University Press, 1986), 194. Italics in original.
85 Kirk Mariner, *Once Upon an Island: The History of Chincoteague* (New Church: Miona Publications, 1996), 105.
86 "Penning the Ponies of Chincoteague," *The Sun* (Baltimore), September 17, 1911, L4.
87 Ralph Pool, "An Isle of Ponies," *The Sun* (Baltimore), August 11, 1929, SM6.
88 "Hundreds of Ponies on Chincoteague Know Only One Day When They are Restrained," *Philadelphia Inquirer*, July 27, 1913, 1.
89 "Penning the Ponies of Chincoteague," *The Sun* (Baltimore), September 17, 1911, L4.
90 "Motor Vehicle Registrations, by States, 1900–1995," Federal Highway Administration, U. S. Department of Transportation. URL: www.fhwa.dot.gov/.
91 "Chincoteague has Road to Mainland," *Baltimore American*, November 16, 1922, 4; Mariner, *Once Upon an Island*, 113.
92 National Archives (College Park, MD), Record Group 16 – Records of the Office of the Secretary of Agriculture, 1906–70 / Folder – Animals-Horses / "Average Prices to Producers in Virginia: Horses."

93  Mariner, *Once Upon an Island*, 117.
94  Ralph Pool, "An Isle of Ponies," *The Sun* (Baltimore), August 11, 1929, SM6.
95  Mariner, *Once Upon an Island*, 118.
96  Ben H. Miller, "Chincoteague's Round-Up of Wild Horses," *The Sun* (Baltimore), July 23, 1933, 4.
97  Marshall Andrews, "Chincoteague Fishermen Turn Cowboy for Annual Pony Drive," *Washington Post*, July 28, 1938, 7.
98  Fred A. Olds, "The Wild Horse of the Banks," *Forest and Stream*, November 15, 1902, 384.
99  "Natural History: North Carolina Wild Horses," *Forest and Stream*, October 17, 1903, 297.
100  Carmine Prioli, *The Wild Horses of Shackleford Banks* (Winston-Salem: John F. Blair, 2007), 44.
101  Chater, "Motor-Coaching through North Carolina," 482.
102  Chater, "Motor-Coaching through North Carolina," 479.
103  "Progress May Mean End of Wild Ponies," *Raleigh News and Observer*, May 5, 1935, 1.
104  "Dare Sharpshooters Begin Banker Ponies' Extinction," *Raleigh News and Observer*, June 14, 1938, 1. See also: Federal Writers Project, *North Carolina: A Guide to the Old North State* (Raleigh: North Carolina Department of Conservation and Development, 1939), 301.
105  Quoted in Aycock Brown, "Ocracoke's Banker Ponies Face Possible Destruction," *Raleigh News & Observer*, June 20, 1938.
106  Mary R. Bullard, *Cumberland Island: a History* (Athens: University of Georgia Press, 2005), 228.
107  "Relief Forces to Corral Wild Horses of Florida," *New York Times*, October 29, 1934, 36.
108  "Uncle Sam: Horse Trader," *Washington Post*, December 6, 1934, 8.
109  "Wild Horses will be Captured," *The Miami News*, October 29, 1934, 5.
110  "Ruth Parker to Scholz, 29 October 1934," in *Looking for the New Deal: Florida Women's Letters during the Great Depression*, edited by Elna C. Green (Columbia: University of South Carolina Press, 2007), 85.
111  R. Dewitt Ivey, "The Mammals of Palm Valley, Florida," *Journal of Mammalogy* 40 (1959): 590.
112  Harriet Ritvo, "Animal Planet," *Environmental History* 9 2 (April, 2004): 215.
113  Harriet Ritvo, "Pride and Pedigree: The Evolution of the Victorian Dog Fancy," *Victorian Studies* 29 (Winter, 1986): 240; Harriet Ritvo, *The Animal Estate: The English and Other Creatures in the Victorian Age* (Cambridge: Harvard University Press, 1987), 240.
114  Katherine C. Grier, *Pets in America: A History* (Chapel Hill: University of North Carolina Press, 2006), 37; Jessica Wang, "Dogs and the Making of the American State," *Journal of American History* 98 (March, 2012): 998–1024.
115  "20th Century Statistics," *Statistical Abstract of the United States* (Washington, DC: U.S. Census Bureau, 1999), 871.
116  Wanda Rushing and Charles Reagan Wilson, *The New Encyclopedia of Southern Culture*: vol. 15: *Urbanization* (Chapel Hill: University of North Carolina Press, 2010); Kirby, *Rural Worlds Lost*, 275.

117 To be sure, there was industrialization in the South prior to the Civil War. John Majewski, *Modernizing a Slave Economy: The Economic Vision of the Confederate Nation* (Chapel Hill; University of North Carolina Press, 2014); Michael J. Gagnon, *Transition to an Industrial South: Athens, Georgia, 1830–1870* (Baton Rouge: Louisiana State University Press, 2012).
118 Kirby, *Rural Worlds Lost*, 286.
119 William A. Link, *Atlanta, Cradle of the New South: Race and Remembering in the Civil War's Aftermath* (Chapel Hill: University of North Carolina Press, 2013).
120 Quoted in Franklin M. Garrett, *Atlanta and Environs: a Chronicle of its People and Events, 1880s–1930s* (Athens: University of Georgia Press, 2011), 263.
121 Robert L. Adamson, "Talk about Dogs," *Atlanta Constitution*, January 3, 1892, 2.
122 "Kennel Club is Formed," *Atlanta Constitution*, January 6, 1900, 9; "Many Find Dogs for Bench Show," *Atlanta Constitution*, May 6, 1900, 7; "Bench Show of the Atlanta Kennel Club to be in Full Blast this Morning," *Atlanta Constitution*, May 9, 1900, 5; "South's Greatest Bench Show Opened," *Atlanta Constitution*, May 10, 1900, 1.
123 "Bench Show of the Atlanta Kennel Club…" *Atlanta Constitution*, May 9, 1900, 5.
124 Sarge Plunkett, "Watching the Procession on Memorial Day Brings Up Memories," *Atlanta Constitution*, May 5, 1895, 2; Sarge Plunkett, "'Possums in Emory Woods," *Atlanta Constitution*, November 2, 1919, D2; "Plunkett Talks of War," *Atlanta Constitution*, October 7, 1906, F9; "The Dog Cart," *The Daily Constitutionalist*, July 18, 1879, 4. Note that other places in the South also reported a dramatic increase in "wild dogs" in the wake of the Civil War. Throughout the 1870s, Fairfax County in northern Virginia directed enormous resources toward combating free-ranging dogs. Albert E. Cowdrey, "Environments of War," *Environmental Review* 7 (summer, 1983): 162.
125 "Where's the Dog Wagon?" *Atlanta Constitution*, June 27, 1892, 5; "War on the Dogs," *Atlanta Constitution*, April 9, 1896, 8; "Mayor Signs the Dog Law," *Atlanta Constitution*, September 23, 1897, 7.
126 *Abstract of the Twelfth Census of the United States, 1900* (Washington, DC: U.S. Government Printing Office, 1904), 40–41; Russell Thornton, "Native American Demographic and Tribal Survival into the Twenty-first Century," *American Studies* 46 (fall/winter, 2005): 23–38.
127 James N. Gregory, *The Southern Diaspora: How the Great Migrations of Black and White Southerners Transformed America* (Chapel Hill: University of North Carolina Press, 2005; Isabel Wilkerson, *The Warmth of Other Suns: The Epic Story of America's Great Migration* (New York:Knopf, 2010); Louis Kyriakoudes, "Rural-Urban Migration," *New Encyclopedia of Southern Culture*: vol. 11: *Agriculture and Industry*, edited by Melissa Walker and James C. Cobb (Chapel Hill: University of North Carolina Press, 2008), 117–120. Regarding urbanization west of the Appalachians, see: Louis Kyriakoudes, *Social Origins of the Urban South: Race, Gender, and Migration in Nashville and Middle Tennessee, 1890–1930* (Chapel Hill: University of North Carolina Press, 2004).

128  Edward A. Hatfield, "Atlanta, Georgia," *Encyclopedia of Southern Culture*, vol. 15 – *Urbanization*, edited by Wanda Rushing (Chapel Hill: University Of North Carolina, 2010), 178.
129  "The Dog Cart," *The Daily Constitutionalist*, July 18, 1879, 4; "Stray Dogs Stampede" *Atlanta Constitution*, October 7, 1925, 24; "Almost a Riot over a Cur Dog," *Atlanta Constitution*, August 27, 1899, 15.
130  Innumerable sources linked feral dogs and rabies. For a sampling, see "Where's the Dog Wagon?" *Atlanta Constitution*, June 27, 1892, 5; "Dog Catcher Now About," *Charleston News and Courier*, June 2, 1908, 10; "Raise Doubt of Rabies," *Washington Post*, December 25, 1911, 5; "99 Dogs are Found Infected with Rabies," *Atlanta Constitution*, January 21, 1936, 2; B. B. Glove, "A Few Facts about the Disease of Rabies," *The Lexington Gazette*, January 4, 1911, 1; "Board of Health Articles," *Ocala Evening Star*, January 27, 1909, 1; "Muzzle the Dogs," *The Times and Democrat* (Orangeburg, SC), July 1910, 4; John E. Watkins, "The Mystery of Rabies: Timely Topics for the Dog Days," *Times-Dispatch* (Richmond), July 14, 1912. For more on the history of rabid dogs in different contexts, see: Bill Wasik and Monica Murphy, *Rabid: A Cultural History of the World's Most Diabolical Virus* (New York: Penguin, 2012); Philip M. Teigen, "The Global History of Rabies and the Historian's Gaze," *Journal of the History of Medicine and Allied Sciences* 67 (2012): 318–327; Wang, "Dogs and the Making of the American State," 998–1024; Karen Brown, "Rabid Epidemiologies: The Emergence and Resurgence of Rabies in Twentieth Century South Africa," *Journal of the History of Biology* 44 (2011): 81–101; Neil Pemberton and Michael Worboys, *Mad Dogs and Englishmen: Rabies in Britain, 1830–2000* (New York: Palgrave Macmillan, 2007); Bert Hansen, "America's First Medical Breakthrough: How Popular Excitement about a French Rabies Cure in 1885 Raised New Expectations for Medical Progress," *American Historical Review* 103 (1998): 373–418.
131  "Darktown's Dogs Die," *Atlanta Georgian and News*, May 29, 1908, 3.
132  "In Olden Times," *Atlanta Constitution*, July 12, 1901, 10; "Poisoned Meat was Used," *Atlanta Constitution*, September 11, 1900, 11; "Dog Catcher is Merciful," *Atlanta Constitution*, May 16, 1903, 7.
133  "Stray Dogs to be Caught by Humane Society," *Atlanta Constitution*, June 20, 1908, 7; "Electricity will be used to Kill Wandering Dogs," *Atlanta Constitution*, May 11, 1921, 10; "Ownerless Dogs and Cats to Die By Electrocution," *Atlanta Constitution*, May 22, 1921, a7.
134  "Just Like Constantinople," *Columbus Ledger* (GA), June 16, 1903, 2. "Dog Wagon on Night Duty," *Atlanta Constitution*, May 21, 1903, 6; "2,000 Dogs Have Been Impounded In Eight Weeks," *Atlanta Constitution*, September 12, 1920, 2a; "City Renews War on Stray Canines," *Atlanta Constitution*, July 23, 1935, 12; "Soft-Hearted Dog Catchers Have a Way," *Atlanta Constitution*, December 12, 1938, 1; "Owners in Rush To Reclaim Pets," *Atlanta Constitution*, December 13, 1938, 4.
135  "The Kennel: Richmond Dog Show," *Forest and Stream* 11 (January 30, 1879): 533; "Letter 15," *Forest and Stream* 31 (September 6, 1888): 130; "Letter 14," *Forest and Stream* 31 (October 25, 1888): 270.

136 "Charleston Dog Show," *Forest and Stream* 38 (January 14, 1892): 35.
137 R. W. Simpson, "Kidnaping [sic] Thoroughbred Dogs," *The Sunny South*, June 1, 1901, 3.
138 "Dog Show added to big Poultry Event," *Charlotte Observer*, August 11, 1916, 10; "Dog Show at the Fair," News and Courier (Columbus GA), July 14, 1909, 5; "Kennel Club's First Dog Show this Fall," *Augusta Chronicle*, October 12, 1913; "Dog Show at Camden," *The State* (Columbia SC), February 10, 1916, 7; "Second Annual Dog Show at Pinehurst," *Charlotte Sunday Observer*, March 9, 1919, 10; "Dog Show at the Fair," *News and Courier* (Columbus GA), July 14, 1909, 5; *Dog Weekly*, vol. 10, no. 1, March 15, 1913, 163. Archives Center, National Museum of American History, Washington DC / Collection no. 60 – Warshaw Collection of Business Americana / Box no. 4 – animals / Folder – Dog Weekly.
139 "The Dog has had his day here, says the law," *The Times* (Richmond), December 1, 1901, 19; "The Plague of Dogs," *Richmond Dispatch*, July 21, 1901, 4. A Dog Fancier, "Dogs at the Pound and the City Hall," *Richmond Times-Dispatch*, August 10, 1902, 5.
140 "Dog Nuisance Still Unabated," *Evening Post* (Charleston SC), October 6, 1902, 5; "Vagrant Dogs Roam at Large," *Evening Post* (Charleston SC), July 26, 1905, 5; "Dog Catcher Now About," *Charleston News and Courier*, June 2, 1908, 10.
141 "Away with the Dogs," *Tampa Morning Tribune*, August 20, 1911, 12.
142 "Regardless of Facts," *Greensboro Record*, July 26, 1918, 4; "Dog Control," *Greensboro Record*, August 1, 1918, 4; "Advertised Curs as High-Bred Dogs," *Richmond Times-Dispatch*, April 6, 1905, 9.
143 "The Dog Nuisance," *Richmond Times Dispatch*, March 5, 1906, 4; "90,313 Dogs," *Charlotte Observer*, December 18, 1905, 8; "Stop Stray Dogs," *Atlanta Constitution*, April 8, 1919, 4; "Macon Dog Catcher Hit by 'Hard Times,'" *Macon Telegraph*, April 11, 1915, 9; "Shooting the Dogs," *Greensboro Record*, June 23, 1916, 8; "Active Workers Trap Many Dogs," *The State* (Columbia SC), July 23, 1919, 10.
144 "Cleaning Out a Den of Wild Dogs," *Atlanta Constitution*, May 17, 1887, 4.
145 "The Day in Georgia," *Atlanta Constitution*, October 19, 1888, 2.
146 "State News Notes," *Atlanta Constitution*, November 4, 1895, 4.
147 "North Carolina," *Baltimore Sun*, August 19, 1899, 8.
148 "Wild Dogs Kill Stock," *Washington Post*, August 16, 1907, 9.
149 Fred A. Olds, "Wild Dogs and a Talking 'Possum," *Forest and Stream*, August 31, 1912, 267; Fred A. Olds, "Wild Dogs and Rabbit," *Forest and Stream*, March 13, 1909, 416. For other examples of feral dogs in the rural South, see "Wild Dogs Decimate Virginia Sheep Flock," *Washington Post*, November 4, 1928, m2; "Pack of Wild Dogs Menaces Game and Pets near Capital," *Washington Post*, March 20, 1935, 30; "Semi-Wild Canines Held Responsible in Raids," *Washington Post*, January 5, 1936, 11; "Fisherman's Luck," *Baltimore Sun*, September 6, 1942, 7.

## 6 Everything in its Right Place

1 Frank Hobbs and Nicole Stoops, *Demographic Trends in the 20th Century* (Washington, DC: United States Census Bureau (November, 2002).
2 Stephanie A. Bohon and Meghan Conley, "Population Change and Demographics," in *New Encyclopedia of Southern Culture: Volume 15 – Urbanization* (Chapel Hill: University of North Carolina Press, 2010), 129; Susanna Robbins, "Keeping Things Cool: Air-Conditioning in the Modern World," *OAH Magazine of History* 18 (October, 2003): 42–46; Raymond Arsenault, "The End of the Long Hot Summer: The Air Conditioner and Southern Culture," *Journal of Southern History* 50 (1984): 597–628.
3 Theda Purdue, "Legacy of Indian Removal" *Journal of Southern History* 78 (2012): 3–36.
4 Regarding biological and/or cultural admixture in the South, see: Leon Fink, "New People of the Newest South: Prospects for the Post-1980 Immigrants," *Journal of Southern History* 75 (August, 2009): 739–750; Carl L. Bankston III, "New People in the New South: An Overview of Southern Immigration," *Southern Cultures* 13 (2007): 24–44; Patrick L. Peacock, Harry L. Watson, and Carrie R. Matthews (eds.), *The American South in a Global World* (Chapel Hill: University of North Carolina Press, 2005); James C. Cobb and William Struek (eds.), *Globalization and the American South* (Athens: University of Georgia Press, 2005); "Table 3: Population for Selected Categories of Race," *2010 United States Census: Summary Population and Housing Characteristics* (January, 2013): 6. For more on Native Americans and racial admixture, see: Russell Thornton, "Native American Demographic and Tribal Survival into the Twenty-First Century," *American Studies* 46 (fall/winter, 2005), 25, 27; David A. Hollinger, "Amalgamation and Hypodescent: The Question of Ethnoracial Mixture in the History of the United States," *American Historical Review* 108 (December, 2003): 1363–1390; Russell Thornton, "Population History of Native North Americans," in *A Population History of North America*, edited by Michael R. Haines and Richard H. Steckel (New York: Cambridge University Press, 2000), xx; Garrett Hellenthal, et al., "A Genetic Atlas of Human Admixture History," *Science* 343 (February 14, 2014): 747–751; Jennifer Hochschild and Velsa Mae Weaver, "'There's No One as Irish as Barack O'Bama': The Policy and Politics of American Multiracialism," *Perspectives on Politics* 8 (September, 2010): 737–759; Nicholas A. Jones and Jungmiwha J. Bullock, "Understanding Who Reported Multiple Races in the U. S. Decennial Census: Results from Census 2000 and the 2010 Census," *Family Relations* 62 (February, 2013): 5–16; *Multiracial in America: Proud, Diverse and Growing in Numbers* (Washington, DC: Pew Research Center, 2015); Kararzyna Bryc, et al., "The Genetic Ancestry of African Americans, Latinos, and European Americans Across the United States," *American Journal of Human Genetics* 96 (January 8, 2015): 37–53; Iosif Lazaridis, et al., "Ancient Human Genomes Suggest Three Ancestral Populations for Present-Day Europeans," *Nature* 513 (September 18, 2014): 409–413; Jared Diamond, "Human Melting Pots in Southeast Asia," *Nature* 262 (August 21, 2014): 262–263; George B. J.

Busby, et al., "The Role of Recent Admixture in Forming the Contemporary West Eurasian Genomic Landscape," *Current Biology* 25 (2015): 1–9.

5 On those occasions when mules are counted, they are invariably lumped with other equids, like horses and burros (donkeys). "Table 18: Horses and Mules," 1959 Agricultural Census, vol. 2, part 6, 508; George B. Ellenberg, "African Americans, Mules, and the Southern Mindscape, 1850–1950," *Agricultural History* 72 (spring, 1998): 382; for the Herotodus quotation, see Albert C. Leighton, "The Mule as a Cultural Invention," *Technology and Culture* 8 (January, 1967): 47.

6 "Horses, Mules, and Motor Vehicles," United States Department of Agriculture Statistical Bulletin no. 5 (Washington, DC: US Government Printing Office, January, 1925), 2: located in RG 17 – Bureau of Animal Industry / Division of Animal Husbandry / Records Pertaining to Horses and Dogs – Box 6, National Archives, College Park, Maryland; *1974 Census of Agriculture*, vol. 1, part 51, US Department of Commerce, Bureau of the Census (December 1977), II-32.

7 Paul G. Irwin, "Overview: the State of Animals in 2001," *The State of the Animals, 2001*, edited by Deborah J. Salem and Andrew N. Rowan (Washington, DC: Humane Society Press, 2001), 8; *2007 Census of Agriculture, State Data*, United States Department of Agriculture, National Agricultural Statistics Service, 426.

8 Emily R. Kilby, "The Demographics of the U. S. Equine Population," in *The State of the Animals*, edited by Deborah J. Salem and Andrew N. Rowan (Washington, DC: Humane Society Press, 2007), 194.

9 Charlene R. Johnson, *Central Florida Thoroughbreds: A History of Horses in the Heart of Florida* (Charleston: The History Press, 2014).

10 "The Kentucky Derby," *Sports Illustrated*, May 7, 1956.

11 Charlene R. Johnson, *Central Florida Thoroughbreds: A History of Horses in the Heart of Florida* (Charleston: The History Press, 2014); Brian Tyrrell, "Bred for the Race," *Historical Studies in the Natural Sciences* 45 (September, 2015): 549–576.

12 Kate Chenery Tweedy, et al., *Secretariat's Meadow: The Land, the Family, the Legend* (Manakin-Sabot: Dementi Milestone Publishing, 2010).

13 Mim A. Bower, et al., "The Genetic Origin and History of Speed in the Thoroughbred Racehorse," *Nature Communications* 2011 (July 27, 2011): 643.

14 Mikkel Schubert, et al., "Prehistoric genomes reveal the genetic foundation and cost of horse domestication," *Proceedings of the National Academy of Sciences* (December 15, 2014): e5661–e5669; Ann Gibbons, "The Thoroughly Bred Horse," *Science* 346 (December, 2014): 1439.

15 "Table 18: Equine," *2012 Census of Agriculture – vol. 1, ch.r 2 – State Data*, 375. Data from other sources suggest much the same. The American Veterinary Medical Association's (AVMA) latest census indicates that the nation's horse population experienced a 31-percent decline between 2011 and 2006. In similar fashion, the American Pet Products Association's (APPA) latest census suggests that the nation's horse population experienced a 41-percent decline between 2010 and 2008. Neither the AVMA nor the APPA provided state-by-state breakdowns. The American Horse Council published the results of its most recent census in 2005. *The Economic Impact of the Horse Industry on the*

United States, American Horse Council Foundation (2005). See also: Kilby, "The Demographics of the U. S. Equine Population"; *APPA National Pet Owners Survey, 2011–2012,* 9, 531.

16. "Horse Welfare: Action Needed to Address Unintended Consequences of Domestic Slaughter," United States Government Accountability Office (June, 2011), i.
17. "Horse Welfare: Action Needed to Address Unintended Consequences from Cessation of Domestic Slaughter," United States Government Accountability Office, Report to Congressional Committees, (June, 2011).
18. *U. S. Pet Ownership & Demographics Sourcebook* (Schaumburg, IL: American Veterinary Medical Association, 2012), 44; *APPA National Pet Owners Survey, 2011–2012* (Greenwich, CTonnecticut: American Pet Products Association, 2012), 7.
19. Marguerite Henry, *Misty of Chincoteague* (Chicago: Rand McNally, 1947).
20. Ashbel Green, "Future to Bridge Virginia: Peninsula Divided 50,000 Visitors," *Christian Science Monitor,* January 26, 1962, 5.
21. Ronald Keiper, *The Assateague Horses* (Tidewater Publishers, 1985), 17.
22. Lori S. Eggert, et al., "Pedigrees and the Study of the Wild Horse Population of Assateague Island National Seashore," *Journal of Wildlife Management* 74 (2010): 963.
23. Keiper, *Assateague Horses,* 18, 91.
24. Eggert, et al., "Pedigrees and the Study of the Wild Horse Population of Assateague Island National Seashore," 963, 970–971.
25. F. G. S., "Beach and other Ponies," *Turf, Field, and Farm* 43 (August 13, 1886), 154; Marinus James, "A Little Journey To Assateague," *Peninsula Enterprise* (September 28, 1929); Clyde DuBose, "Chincoteague Island Ready for Round-Up of Wild Ponies," *Washington Post,* July 22, 1934, B2; Gerald B. Webb, Jr. "Sea Horses of Chincoteague," *The Spur* 62 (August 1, 1938), 36.
26. Marguerite Henry, *Stormy, Misty's Foal* (Chicago: Rand McNally, 1963).
27. Keiper, *Assateague Ponies,* 18–19.
28. Louise Sweeney, "Chincoteague ponies face ordeal," *Baltimore Sun,* July 23, 1978, h2.
29. Phone conversation with Denise Bowden, Chincoteague Volunteer Fire Department, Chincoteague, Virginia, October 11, 2013.
30. The gift shop sells a book that insists the Assateague horses owe their origin to a shipwreck. John Amrhein, Jr., *The Hidden Galleon: The True Story of a Lost Spanish Ship and the Legendary Wild Horses of Assateague Island* (Kitty Hawk: New Maritima Press, 2007).
31. "Karl T. Gilbert (superintendent) letter to Thomas C. Ellis (Superintendent of State Parks), April 1, 1965." State Archives of North Carolina / Old Records Center (W1-C) / State Parks Division: Miscellaneous Records / Folder – Ponies Ocracoke, General.
32. Bonnie Urquhart Gruenberg, *Hoofprints in the Sand: Wild Horses of the Atlantic Coast* (Strasburg: Eclipse Press, 2002), 75.
33. "Article 4: Stock along the Outer Banks." State Archives of North Carolina / Old Records Center (W1-C) / State Parks Division: Miscellaneous Records / Folder – Ponies Ocracoke, General. Raleigh.

34 "Park Service Probably to Maintain Pony Herd," *Raleigh News and Observer,* June 19, 1966. Vertical File: Ponies in North Caro." State Archives of North Carolina. Raleigh.
35 "Park Service Probably to Maintain Pony Herd," *Raleigh News and Observer,* June 19, 1966. Vertical File: Ponies in North Caro." State Archives of North Carolina. Raleigh.
36 "Karl T. Gilbert (superintendent) letter to Thomas C. Ellis (Superintendent of State Parks), April 1, 1965." State Archives of North Carolina / Old Records Center (W1-C) / State Parks Division: Miscellaneous Records / Folder – Ponies Ocracoke, General.
37 Urquhart, *Hoofprints in the Sand,* 75, 80–82.
38 Carmine Prioli and Scott Taylor, *The Wild Horses of Shackleford Banks* (Winston Salem: John F. Blair Publishing, 2007), 12, 100; E. K. Conant, R. Juras and E. G. Cothran, "A Microsatellite Analysis of Five Colonial Spanish Horse Populations of the Southeastern United States," *Animal Genetics* 43 (February, 2012): 54.
39 Prioli and Taylor, *The Wild Horses of Shackleford Banks,* 44; Old Trudge, "Penning Ponies," *The State,* July 5, 1952. Vertical File: Ponies in North Caro." State Archives of North Carolina. Raleigh.
40 Norwood Young, "The Pony is Disappearing," *News and Observer,* August 8, 1954. Vertical File: Ponies in North Caro." State Archives of North Carolina. Raleigh.
41 "Chapter 1057, Session Laws 1957: An Act to Prohibit Stock and Cattle from Running at Large Along the Outer Banks." June 5, 1957. State Archives of North Carolina / Old Records Center (W1-C) / State Parks Division: Miscellaneous Records / Folder – Ponies Ocracoke, General.
42 "Letter from Thomas W. Morse to Robert E. Giles, July 27, 1960." State Archives of North Carolina / Old Records Center (W1-C) / State Parks Division: Miscellaneous Records / Folder – Ponies Ocracoke, General.
43 Conant et al., "A Microsatellite Analysis of Five Colonial Spanish horse populations," 59.
44 Urquhart, *Hoofprints in the Sand,* 52.
45 "An Act to Ensure Maintenance of a Herd of Wild Horses, January 27, 1998," reprinted in Prioli and Taylor, *The Wild Horses of Shackleford Banks,* 70–71.
46 Prioli and Taylor, *The Wild Horses of Shackleford Banks,* 12, 100; Conant et al., "A Microsatellite Analysis of Five Colonial Spanish horse populations," 54.
47 Karen Hileman McCalpin, *Saving the Horses of Kings: The Wild Horses of Currituck Outer Banks* (Kitty Hawk: Outer Banks Press, 2010); Urquhart, *Hoofprints in the Sand,* 36.
48 Urquhart, *Hoofprints in the Sand,* 36.
49 Fran Lynhaug, *The Official Horse Breeds Standard Guide* (Voyageur, 2009), 37.
50 Conant et al., "A Microsatellite Analysis of Five Colonial Spanish horse populations," 61; Kimberly M. Porter, "Vegetative Impact of Feral Horses, Feral Pigs, and Whitetailed Deer on the Currituck National Wildlife Refuge, North Carolina," *Castanea* 79 (2014): 8–17; D. P. Sponenberger, "Deciding Which Feral Horse Populations Qualify as a Genetic Resource," *Bureau of Land Management Resource Notes* 25 (July 19, 2000); Porter, et al., "Vegetative

Impact of Feral Horses, Feral Pigs, and White-tailed Deer on the Currituck National Wildlife Refuge, North Carolina," 8–17.

51 "Cumberland Island Feral Horse Facts," document provided by Doug Hoffman, biologist at Cumberland Island National Seashore, in an email to the author. December 11, 2012.

52 In 2011, the Park's Livestock Management Plan recommended stabilizing the population at current levels. In theory, the park could put surplus horses up for adoption, but officials point out that there was "not a large demand for unbroken horses." Instead, they recommend maintaining a herd that consists of females and gelded (neutered) males only. This would allow them to breed and feralize additional horses on an as-needed basis. "Livestock Management Plan: Paynes Prairie Preserve State Park," State of Florida, Department of Environmental Protection, Division of Recreation and Parks (March 2011), 13–14; "Florida Cracker Horse," *The Official Horse Breeds Standard Guide* (2012), 76.

53 Lynhaug, *Horse Breeds Standard Guide*, 47.

54 Jeff Kidd, "Update: Marsh ponies of Little Horse Island make it through difficult winter," *Island Packet* (Bluffton, SC), March 9, 2015, 1.

55 Grayson Highlands State Park website. URL: www.dcr.virginia.gov/stateparks/grayson-highlands.shtml.

56 Folder: Horse, Przewaski / Box 23 / Series 4 / RU 365 – National Zoological Park, Office of Public Affairs, 1899–1988 / Smithsonian Institution Archives, Washington, DC.

57 Canjun Xia, et al., "Reintroduction of Przewalski's horse (*Equus ferus przewalskii*) in Xinjiang, China: The Status and Experience," *Biological Conservation* 177 (2014): 142.

58 Lee Boyd and Katherine A. Houpt, *Przewalski's Horse: The History and Biology of an Endangered Species* (Albany: State University of New York Press, 1994); Emily Shenk, "First Przewalski's Horse Born via Artificial Insemination," *National Geographic News*, August 5, 2013.

59 Perry Van Ewing, *Southern Pork Production* (New York: Orange Judd Company, 1918), 181.

60 Michael D. Thompson, "High on the Hog: Swine as Culture and Commodity in Eastern North Carolina," PhD Dissertation, Miami University, 2000, 72.

61 Jack Temple Kirby, *Mockingbird Song: Ecological Landscapes of the South* (Chapel Hill: University of North Carolina, 2006), 197.

62 *1964 Census of Agriculture*, vol. 1, part 28, 12; part 26, 12; Michael D. Thompson, "This Little Piggy Went to Market: The Commercialization of Hog Production in Eastern North Carolina from William Shay to Wendell Murphy," *Agricultural History* 74 (spring, 2000): 575.

63 Thompson, "High on the Hog," 74, 164.

64 Monica Gisolfi, "From Crop Lien to Contract Farming: The Roots of Agribusiness in the American South," *Agricultural History* 80 (spring, 2006): 167–189; Thompson, "High on the Hog," 168; William Boyd, "Making Meat: Science, Technology, and American Poultry Production," *Technology and Culture*, 42 (October, 2001): 631–664; Charles R. Wilson, "Agribusiness," in *New Encyclopedia of Southern Culture, Volume 11: Agriculture*

*and Industry*, edited by Melissa Walker and James C. Cobb (Chapel Hill: University of North Carolina Press, 2008); Deborah Fitzgerald, *Every Farm a Factory: The Industrial Ideal in American Agriculture* (New Haven: Yale University Press, 2003).

65 Thompson, "High on the Hog," 166–167; Thompson, "This Little Piggy Went to Market," 578–579.
66 Thompson, "High on the Hog," 75, 78–79; Thompson, "This Little Piggy Went to Market," 569.
67 *1964 Census of Agriculture*, vol. 2, part 2, (Washington, DC: U. S. Bureau of the Census, 1967), 44; *1987 Census of Agriculture – United States Data*, National Agricultural Statistics Service, p. 7; *1964 Census of Agriculture*, vol. 2, part 2, 44; *2007 Census of Agriculture – United States Data*, National Agricultural Statistics Service, 7.
68 "Overview of the U. S. Hog Industry," National Agricultural Statistics Service, (October 30, 2009), 1. See also: Joel Salatin, *Folks, This Ain't Normal: A Farmer's Advice for Happier Hens, Healthier People, and a Better World* (New York: Center Street, 2012).
69 Hongjun Tao and Chaoping Xie, "A Case Study of Shuanghui International's Strategic Acquisition of Smithfield Foods," *International Food and Agribusiness Management Review* 18 (2015), 148; Marion G. Everett and Patsy D. Barham, *A History of Smithfield Ham Industry* (Smithfield: Isle of Wight County Museum, 1993); Patrick Evans-Hylton, *Smithfield: Ham Capital of the World* (Charleston: Arcadia, 2004).
70 "Chinese Company Buys Smithfield Foods Inc. in $4.72 billion deal," *Fayetteville Observer*, May 30, 2013.
71 "North Carolina," *2007 Census of Agriculture* (Washington, DC: National Agricultural Statistics Service), 431–443.
72 Cody Carlson, "How State Ag-Gag Laws Could Stop Animal-Cruelty Whistleblowers," *The Atlantic*, March 25, 2013.
73 Michael Pollan, *The Omnivore's Dilemma: A Natural History of Four Meals* (New York: Penguin, 2006), 226–239; Timothy Pachirat, *Every Twelve Seconds: Industrialized Slaughter and the Politics of Sight* (New Haven: Yale University Press, 2011).
74 "Overview of the U. S. Hog Industry," National Agricultural Statistics Service, (October 30, 2009), 1.
75 "The State of the World's Animal Genetic Resources for Food and Agriculture" (United Nations Commission on Genetic Resources for Food and Agriculture, 2007), 5.
76 David Cryanoski, "Super-Muscly Pigs Created by Small Genetic Tweak," *Nature* 523 (July 2, 2015): 13–14; Yu-Hsis Su, et al., "Construction of a CRISPR-Cas9 System for Pig Genome Targeting," *Animal Biotechnology* 26 (2015): 279–288; David Cyranoski, "Gene-edited Pigs to be Sold as Pets," *Nature* 527 (October 1, 2015): 18.
77 Tao and Xie, "A Case Study of Shuanghui International's Strategic Acquisition of Smithfield Foods," 149.
78 Edward Wyatt, "Senators Question Chinese Takeover of Smithfield," *New York Times*, July 10, 2013.

79 Michael J. De La Merced, "U. S. Security Panel Clears a Chinese Takeover of Smithfield Foods," *New York Times*, September 6, 2013.
80 Maureen Ogle, *In Meat We Trust: An Unexpected History of Carnivore America* (Boston: Houghton Mifflin, 2013); Christopher G. Davis and Biing-Hwan Lin, "Factors affecting U.S. Pork Consumption," *USDA ERS Electronic Outlook Report* (2005), 1–2; Richard A. Lobban, Jr., "Pigs and their Prohibition," *International Journal of Middle East Studies* 26 (February, 1994): 57–75; Daphne Barak-Erez, *Outlawed Pigs: Law, Religion, and Culture in Israel* (Madison: University of Wisconsin Press, 2007).
81 R. P. Hanson and Lars Karstad, "Feral Swine in the Southeastern United States," *Journal of Wildlife Management* 23 (January, 1959): 66; Tom McKnight, "Feral Animals in Anglo-America," *University of California Publications in Geography* 16 (May 6, 1964): 45.
82 John Mayer and Lehr Brisbin, "Introduction," in *Wild Pigs: Biology, Damage, Control Techniques and Management*, edited by John Mayer and Lehr Brisbin (Aiken: Savannah River National Laboratory, 2009), 2.
83 Mayer and Brisbin, *Wild Pigs of the United States*, 1. See also: Meredith McClure, et al., "Modeling and Mapping the Probability of Occurrence of Invasive Wild Pigs across the Contiguous United States," *PLoS ONE*, August 12, 2015, 1–17.
84 T. S. Palmer, "The Danger of Introducing Noxious Animals and Birds," *Forest and Stream* 52 (June 3, 1899), 425.
85 "The Razorback Hog," *Forest and Stream* 72 (March 19, 1910): 2.
86 LeRoy C. Stegeman, "The European Wild Boar in the Cherokee National Forest, Tennessee," *Journal of Mammalogy* 19 (August, 1938): 282.
87 Charles Elton, *The Ecology of Invasions by Plants and Animals* (Chicago: University of Chicago Press, 1958), 89.
88 McKnight, "Feral Animals in Anglo-America," 43.
89 John J. Mayer, "Overview of Wild Pig Damage," in *Wild Pigs: Biology, Damage, Control Techniques and Management*, edited by John Mayer and Lehr Brisbin (Aiken: Savannah River National Laboratory, 2009), 221–246.
90 Mayer, "Overview of Wild Pig Damage," 221.
91 Will Brantley, "The Pig Report," *Field & Stream* (September, 2015), 78–82.
92 Siddhartha Thakur, et al., "Detection of Clostridium difficile and Salmonella in Feral Swine Population in North Carolina," *Journal of Wildlife Diseases* 47 (2011), 774–776; Richard Engeman, "Making Contact: Rooting Out the Potential Exposure of Commercial Production Swine Facilities to Feral Swine in North Carolina, *EcoHealth* 8 (2011), 76–81.
93 Brantley, "The Pig Report," 82.
94 Doug Hoffman, "Efficacy of Shooting as a Control Method for Feral Hogs," in *Wild Pigs: Biology, Damage, Control Techniques and Management*, edited by John Mayer and Lehr Brisbin (Aiken: Savannah River National Laboratory, 2009), 290; Brantley, "The Pig Report," 78–82.
95 Tyler A. Campbell and David B. Long, "Strawberry-Flavored Baits for Pharmaceutical Delivery to Feral Swine," *Journal of Wildlife Management* 73 (May, 2009): 615–619.
96 Hoffman, "Efficacy of Shooting as a Control Method for Feral Hogs," 301.
97 Mayer and Brisbin, "Introduction," 1.

98 "The Razorback Hog," *Forest and Stream*, lxxiv (March 19, 1910): 2.
99 Carmen McCormack, "Boar-Hunting Now Very Big in Miami," *Washington Post and Times Herald*, April 4, 1954, c10.
100 Blake E. McCann, "Mitochondrial Diversity Supports Multiple Origins for Invasive Pigs," *Journal of Wildlife Management* 78 (2014): 202–213; M. Noelia Barrios-Garcia and Sebastian A. Ballari, "Impact of Wild Boar (*Sus scrofa*) in its Introduced and Native Range: A Review," *Biological Invasions* 14 (2012): 2283–2300; Sarah N. Bevins, et al., "Consequences Associated with the Recent Range Expansion of Nonnative Feral Swine," *BioScience* 64 (April, 2014): 291–299; Barrios-Garcia, "Impact of Wild Boar (*Sus scrofa*) in its Introduced and Native Range: A Review," *Biological Invasions* 14 (2012), 2283–2300.
101 James C. Lewis, "Observations of Pen-Reared European Hogs Released for Stocking," *Journal of Wildlife Management* 30 (October, 1966), 832–835.
102 John J. Mayer, "Taxonomy and History of Wild Pigs in the United States," in *Wild Pigs: Biology, Damage, Control Techniques and Management*, edited by John Mayer and Lehr Brisbin (Aiken: Savannah River National Laboratory, 2009), 8; Philip S. Gipson, Bill Hlavachick, and Tommie Berger, "Range Expansion by Wild Hogs across the Central United States," *Wildlife Society Bulletin* 26 (summer, 1998): 280; Mayer, "Taxonomy and History of Wild Pigs in the United States," 13.
103 Mayer, "Taxonomy and History of Wild Pigs in the United States," 13.
104 Jeffrey Greene, *The Golden-Bristled Boar: Last Ferocious Beast of the Forest* (Charlottesville: University of Virginia Press, 2012); Ian Frazier, "Hogs Wild," *New Yorker*, December 12, 2005, 71–83; Ted Chamberlain, "Photo in the News: Hogzilla is No Hogwash," *National Geographic News*, March 22, 2005.
105 Jay Reeves, "Alabama Officials say 'Monster Pig' hunt was legal," *Daily Times* (Florence, Alabama), June 1, 2007, 2b.
106 Rhonda Shearer, "Alabama's Monster Pig Hoax, One Year Later," ESPN.com, May 2, 2008, URL: espn.go.com/espn/print?id=3378412. See also: Brian Stickland, "Monster Pig raised on Fruithurst farm, not a wild hog," *Columbus Ledger-Enquirer*, June 1, 2007.
107 Edmund Russell, *Evolutionary History: Uniting History and Biology to Understand Life on Earth* (New York: Cambridge University Press, 2011), 17.
108 Brantley, "The Pig Report," 82.
109 I. Lehr Brisbin and Michael S. Sturek, "The Pigs of Ossabaw Island: A Case Study of the Application of Long-term Data in Management Plan Development," in *Wild Pigs: Biology, Damage, Control Techniques and Management*, edited by John Mayer and Lehr Brisbin (Aiken: Savannah River National Laboratory, 2009), 365–378.
110 Mart A. Stewart, "'Whether Wast, Deodand, or Stray': Cattle, Culture, and the Environment in Early Georgia," *Agricultural History* 65 (1991): 16.
111 "Offers Island as Hospital," *Atlanta Constitution*, July 11, 1898, 7.
112 D. Phillip Sponenberg, Jeannette, Beranger, and Alison Martin, *An Introduction to Heritage Breeds: Saving and Raising Rare-Breed Livestock and Poultry* (North Adams: Storey Publishing, 2014).

113 Brisbin and Sturek, "The Pigs of Ossabaw Island," in *Wild Pigs: Biology, Damage, Control Techniques and Management*, edited by John Mayer and Lehr Brisbin (Aiken: Savannah River National Laboratory, 2009); I. Lehr Brisbin and John J. Mayer, "Problem Pigs in a Poke: A Good Pool of Data," *Science* 294 (November 9, 2011): 1280–1281.

114 Ed Crews, "Ossabaw Island Pigs," *Colonial Williamsburg Journal* (2010); Scott Magelssen, "Resuscitating the Extinct: The Backbreeding of Historic Animals at U.S. Living History Museums," *The Drama Review* 47 (2003), 98–109.

115 Mayer and Brisbin, "Introduction," 12.

116 While just about everyone loves dogs, women are the ones keeping the animals alive. According to one recent survey, females are the primary shoppers for pet products, including food, in most (78 percent) dog-owning households. *APPA National Pet Owners Survey, 2011–2012*, 58.

117 Elizabeth A. Clancy and Andrew Rowan, "Companion Animal Demographics in the United States: a Historical Perspective," *The State of the Animals, 2003*, edited by Andrew Rowan and Deborah Salem (Washington, DC: Humane Society Press, 2003), 9.

118 The same pattern was exhibited on a nationwide scale. In 1996, the nation contained about 53,000,000 dogs. By 2001, that number had increased to more than 61,000,000 and by 2006 the dog population had increased to more than 72,000,000. When the AVMA released the results of its 2011 survey, however, the national population had actually dropped *below* 70,000,000. *U. S. Pet Ownership & Demographics Sourcebook, 2012* (Schaumburg, IL: American Veterinary Medical Association, 2012), 21–24.

119 Louis Kyriakoudes, "Rural-Urban Migration," *New Encyclopedia of Southern Culture*: vol. 11: *Agriculture and Industry*, edited by Melissa Walker and James C. Cobb (Chapel Hill: University of North Carolina Press, 2008), 19; Orville Vernon Burton, "The South as 'Other,' the Southerner as 'Stranger,'" *Journal of Southern History* (2013): 31–32; Pete Daniel, "The Transformation of the Rural South: 1930 to the Present," *Agricultural History* 55 (July, 1981): 231–248; Matthew D. Lassiter and Kevin M. Kruse, "The Bulldozer Revolution: Suburbs and Southern History since World War II," *Journal of Southern History* (2009):

120 Kirby, *Mockingbird Song*, 163.

121 "Florida: 2010," *2010 Census of Population and Housing* (Washington, DC: U.S. Census Bureau, September 2012).

122 *APPA National Pet Owners Survey, 2011–2012*, 70.

123 *U.S. Pet Ownership & Demographics Sourcebook, 2012*, 18; *APPA National Pet Owners Survey, 2011–2012*, 68.

124 Gwyneth Anne Thayer, *Going to the Dogs: Greyhound Racing, Animal Activism, and American Popular Culture* (Lawrence: University Press of Kansas, 2013).

125 Qingjian Zou, et al., "Generation of Gene-Target Dogs using CRISPRCas9 System," *Journal of Molecular Cell Biology* (Advanced Access: October 21, 2015).

126 *U.S. Pet Ownership & Demographics Sourcebook, 2012*, 11.

127 Greger Larson, et al., "Rethinking Dog Domestication by Integrating Genetics, Archaeology, and Biogeography," *Proceedings of the National Academy of Sciences* 109 (June 5, 2012), 8880.
128 Heidi G. Parker, et al., "Genetic Structure of the Purebred Domestic Dog," *Science* 304 (May 21, 2004): 1164.
129 Larson et al., "Rethinking Dog Domestication by Integrating Genetics, Archaeology, and Biogeography," 8880.
130 Michael Brandow, *A Matter of Breeding: A Biting History of Pedigree Dogs and How the Quest for Status Has Harmed Man's Best Friend* (Boston: Beacon, 2015); Austin L. Hughes, "Accumulation of Slightly Deleterious Mutations in the Mitochondrial Genome: A Hallmark of Animal Domestication," *Gene* 515 (2013): 28–33; Fernando Cruz, Carles Vila, and Matthew T. Webster, "The Legacy of Domestication: Accumulation of Deleterious Mutations in the Dog Genome" *Molecular Biology and Evolution* 25 (2008): 2331–2336. See also: Harold Herzog, "Forty-Two Thousand and One Dalmatians: Fads, Social Contagion, and Dog Breed Popularity," *Society & Animals* 14 (2006): 383–397; Anthony King, "Domestication Set Dog Genes Free" *New Scientist* 191 (July 7, 2006), 18.
131 In 1964, biogeographer Tom McKnight claimed that there were between 40,000,000 and 50,000,000 dogs in the United States. He did not indicate how many of these animals were feral, but his comments suggest that it was a large percentage. McKnight, "Feral Animals in Anglo-America," 46–48.
132 Carl Djerassi, Andrew Israel, and Wolfgang Jochle, "Planned Parenthood for Pets?" *Bulletin of the Atomic Scientists* 29 (January, 1973): 12.
133 Katherine C. Grier, *Pets in America: A History* (Chapel Hill: University of North Carolina Press, 2006), 101.
134 In 1980, Swedish-born activist Ingrid Newkirk established People for the Ethical Treatment of Animals in Norfolk, Virginia, while renowned NC State philosopher Tom Regan published *In Defense of Animals* in 1983 (University of California Press). Regarding animal-rights advocacy in the South and across the nation, see also: Wayne Pacelle, *The Bond: Our Kinship with Animals, Our Call to Defend Them* (New York: HarperCollins, 2011); Ingrid Newkirk, *The PETA Practical Guide to Animal Rights: Simple Acts of Kindness to Help Animals in Trouble* (New York: St. Martin's Griffin, 2009); Diane L. Beers, *For the Prevention of Cruelty: The History and Legacy of Animal Rights Activism in the United States* (Columbus: Ohio University Press, 2006); Bernard Unti, *Protecting All Animals: A Fifty-Year History of the Humane Society of the United States* (Washington, DC: Humane Society of the United States, 2004); Peter Singer, *Animal Liberation: A New Ethics for Our Treatment of Animals* (New York: Random House, 1975).
135 Clancy and Rowan, "Companion Animal Demographics in the United States,"
136 Clancy and Rowan, "Companion Animal Demographics in the United States," 15, 20; Irwin, "Overview: the State of Animals in 2001," 2.
137 Stephen Coate and Brian Knight, "Pet Overpopulation: an Economic Analysis," *B. E. Journal of Economic Analysis & Policy* 10 (2010): 1; Irwin, "Overview: the State of Animals in 2001," 2.

138 Juan Li, et al., "Vectored Antibody Gene Delivery Mediates Long-Term Contraception," *Current Biology* 25 (October 5, 2015), r820–r822; David Grimm, "A Cure for Euthanasia?" *Science* 325 (September 18, 2009), 1490–1493.
139 Conversation with Xiamora Mordcovich, Miami-Dade Animal Services, spring 2013.
140 Glover Allen, "Dogs of the American Aborigines," *Bulletin of the Museum of Comparative Zoology at Harvard College* 63 (1920): 439.
141 Marison Schwartz, *History of Dogs in the Early Americas* (New Haven: Yale University Press, 1998), 164, 167; Mastromino, "'Cry Havoc and Let Loose the Dogs of War': Canines and the Colonial American Military Experience," Master's Thesis, College of William and Mary, 1986, 25.
142 Santiago Castroviejo-Fisher, et al., "Vanishing American Dog Lineages," *BMC Evolutionary Biology* 11 (2011), 4–5.
143 Conversation with Lehr Brisbin, June 30, 2010. Regarding the evolutionary history of dingoes, see: Bradley P. Smith and Carla A. Litchfield "A Review of the Relationship between Indigenous Australians, Dingoes (*Canis dingo*) and Domestic Dogs (*Canis familiaris*)," *Anthozoos* 22(2, 2009), 111–128.
144 Schwartz, *History of Dogs in the Early Americas*, 213.
145 I. Lehr Brisbin, "Primitive Dogs, their Ecology and Behavior: Unique Opportunities to Study the Early Development of the Human-Canine Bond," *Journal of the American Veterinary Medical Association* 210 (1997), 1122–1126; Scott Weidensaul, "Tracking America's First Dogs," *Smithsonian Magazine* (March 1999).
146 Weidensaul, "Tracking America's First Dogs," 49.
147 David L. White, *Deerskins and Cotton: Ecological Impacts of Historic Land Use in the Central Savannah River Area of the Southeastern U. S. before 1950*, Final Report to the USDA Forest Service, Savannah River (January 2004); Richard David Brooks and David Colin Crass, *A Desperate Poor Country: History and Settlement Patterning on the Savannah River Site*, Savannah River Archaeological Research Papers 2 (University of South Carolina, 1991); Kari Frederickson, *Cold War Dixie: Militarization and Modernization in the American South* (Athens: University of Georgia Press, 2013). For evidence of stray dogs after the establishment of the SRS, see: Louise Cassels, *The Unexpected Exodus: How the Cold War Displaced One Southern Town* (Columbia: University of South Carolina Press, 2007), xxix.
148 Barbara Van Asch et al., "Pre-Columbian Origins of Native American Dog Breeds," *Proceedings of the Royal Society B* 280 (2013): 7.
149 Laura M. Shannon, et al., "Genetic Structure in Village Dogs Reveals a Central Asian Domestication Origin," *Proceedings of the National Academy of Sciences* 112 (November 3, 2015): 13639–13644.
150 Christopher J. Manganiello, "From a Howling Wilderness to Howling Safaris: Science, Policy and Red Wolves in the American South," *Journal of the History of Biology* 42 (2009): 325–359. See also: Philip J. Seddon, "Reversing Defaunation: Restoring Species in a Changing World," *Science* 345 (2014): 406–412; Erik Stokstad, "Red Wolves in the Crosshairs," *Science* 345 (September 26, 2014): 1548.

## Epilogue. Cultivating Ferality in the Anthropocene

1 Others reject that narrative as defeatist, insisting that while human influence is widespread, some wild regions are "still intact." Tim Caro, et al., "Conservation in the Anthropocene," *Conservation Biology* 26 (2011): 185–188. Regarding humanity's pervasive influence, see: William F. Ruddiman, et al., "Defining the Epoch We Live In," *Science* 348 (April 3, 2015): 38–39; Simon L. Lewis and Mark A. Maslin, "Defining the Anthropocene," *Nature* 519 (March 12, 2015): 171–180; Richard Monastersky, "The Human Age," *Nature* 519 (March 12, 2015): 144–147; Jonathan Williams and Paul J. Crutzen, "Perspectives on our Planet in the Anthropocene," *Environmental Chemistry* 10 (2013): 269–280; Eileen Crist, "On the Poverty of Our Nomenclature," *Environmental Humanities* 3 (2013): 129–147; David Oldroyd and Robert Davis, "Inventing the Present: Historical Roots of the Anthropocene," *Earth Sciences History* 30 (2011): 63–84; Will Steffen, Jacques Grinevald, Paul Crutzen and John McNeill, "The Anthropocene: Conceptual and Historical Perspectives," *Philosophical Transactions of the Royal Society A* (January, 2011): 842–867; Will Steffen, et al., "The Anthropocene: From Global Change to Planetary Stewardship," *Ambio* 40 (2011): 739–761; Paul J. Crutzen and Will Steffen, "How Long Have We Been in the Anthropocene Era?" *Climatic Change* 61 (2003): 251–247. Some regard the Anthropocene and the Holocene as one and the same. See: Giacomo Certini and Riccardo Scalengh, "Holocene as Anthropocene," *Science* 349 (July 17, 2015): 246.
2 Gerardo Ceballos, et al., "Accelerated Modern Human-Induced Species Losses: Entering the Sixth Mass Extinction," *Science Advances* 1 (June 19, 2015): e1400253; Douglas J. McCauley, et al., "Marine defaunation: Animal loss in the global ocean," *Science* 347 (January 16, 2015): 247; Rodolfo Dirzo, et al., "Defaunation in the Anthropocene," *Science* 345 (July 25, 2014): 401–406; Elizabeth Kolbert, *The Sixth Extinction: An Unnatural History* (Henry Holt and Company, 2014); S. L. Pimm, et al., "The Biodiversity of Species and Their Rates of Extinction, Distribution, and Protection," *Science* 344 (May 30, 2014): 988.
3 Paul G. Irwin, "Overview: The State of Animals in 2001," in *The State of the Animals, 2001*, edited by Deborah J. Salem and Andrew N. Rowan (Washington, DC: Humane Society Press, 2001), 4.
4 Tom McKnight, *Feral Livestock in Anglo-America* (Berkeley: University of California Press, 1964), 47.
5 Stephen Coate and Brian Knight, "Pet Overpopulation: An Economic Analysis," *B. E. Journal of Economic Analysis & Policy* 10 (2010): 1; Irwin, "Overview: the State of Animals in 2001," 1.

# INDEX

4H, 81

Adams, Charles Francis, 74
Adams, John, 74
Adams, John Quincy, 57
Adamson, Robert L., 104
Africa, 13, 15, 16, 17, 21, 22, 23, 24, 25, 38, 130
ag-gag legislation, 124
agricultural census, 59, 71, 78, 114, 115
air-conditioning, 113
Alabama, 129
Alachua Savannah, 46, 103, 120
Alaska, 14, 28
Albemarle County (Virginia), 57
Albemarle Sound, 38, 39
Alexander the Great, 66
Allen, Glover, 135
Altai Mountains, 121
Amelia Island, 44
American Livestock Breeds Conservancy, 120, 131
*American Turf Register*, 62, 63
American Veterinary Medical Association, 131, 132
Anatolia, 20
Andalusian, 119
Anderson, Virginia DeJohn, 3
Andrews, Marshall, 101
animal studies, 143
Anthropocene, 6, 11, 139, 140
Apalachee Indians, 45
Appalachian Mountains, 6, 8, 27, 59, 60, 66, 71, 78, 88, 92, 128
Appomattox Courthouse, 77
Arabia, 13
Arabian horses, 23, 42, 120
*Arator*, 69

archaeology, 16, 17, 20, 22, 24, 27, 28, 29, 136, 173
Archaic-Era, 169
Argall, Samuel, 33
Argentina, 63
Arizona, 103
Army of Northern Virginia, 76
Asheville, 110
Ashley River, 37
Assateague, 35, 63, 64, 93, 100, 101, 115, 116, 117, 121
Assateague Island National Seashore, 116
Atlanta, 96, 97, 104–8
*Atlanta Constitution*, 104
Atlanta Humane Society, 107
Atlanta Kennel Club, 104
Atlantic Ocean, 6, 8
Aucilla River, 169
Audubon, John James, 170
Augusta, 56, 58, 84, 85, 96, 108
*Augusta Chronicle*, 96, 98
Australia, 14, 17, 24
Austria, 43
automobiles, 95–99, 100, 111, 114, 118, 120
Awash Valley, 13
Ayllón, Lucas Vázquez de, 171

Bab-el-Mandeb Strait, 13
bacon, 67
Bacon, Nathaniel, 36
Bacon's Rebellion, 36
Ball, Charles, 56
Baltimore, 75
*Baltimore Sun*, 101
Barbados, 38
barbecue, 40
Barton, Benjamin Smith, 52, 135

## Index

Bartram, William, 46–49, 103, 120, 170
Belle Island Prison, 74
Belmont Stakes, 114
Belyaev, Dmitry, 18
bench shows, 103, 104, 108, 112
Benjamin, Judah P., 76
Bergmann's Rule, 23
Beringia, 14, 22, 27
Berkeley, William, 36
Berkshire, 87
Bermuda, 32
Beverley, Robert, 37
Bight of Benin, 38
Bight of Biafra, 38
Billie, Cypers, 103
Black Legend, 33
Bladen County (North Carolina), 124
Blissitt, Phil, 129
Blissitt, Rhonda, 129
bloodhounds, 33, 104
Bolzius, John Martin, 44
Borden, Eddy, 129
borderlands, 9
Boston, 93, 107
Botai Culture, 22
*Botany of Desire*, 4
Boy Scouts, 117, 118
*Boy's Life*, 117
Brevard County (Florida), 169
Brickell, John, 41, 42, 43
Brisbin, I. Lehr, 125, 135, 136
Britain, 11, 23, 65, 103
British, 11
Browne, William H., 108
Browne, William M., 84
Bucephalus, 66
Buckingham, James Silk, 62, 66, 67
Buffon, Comte de, 2
Bullock, William, 34
Bureau of Biological Survey, 125
Burke, Emily, 67
Burnaby, Andrew, 42
Burnett, Edmund Cody, 72
Burroughs, William S., 52
Byrd, William, 40

California, 117
Camden (South Carolina), 108
Campbell, John, 55
Canada, 14, 49
Canary Islands, 30

Cantino Planisphere, 171
Cape Hatteras National Seashore, 117, 118
Cape Lookout National Seashore, 119
Caribbean Islands, 29, 30, 34
Carleton, George, 54
Carnegie, Lucy, 95
Carnegie, Thomas, 95
Carnes, H.G., 107
Carolina Dogs, 136
Carolina Jockey Club, 42
Caroline County (Virginia), 69, 114
Castroviejo-Fisher, Santiago, 135
Catawba Indians, 42
Catesby, Mark, 40, 43, 46
cavalry, 75, 76
Charleston, 37, 38, 42, 43, 49, 58, 84, 96, 104, 108, 109, 110
*Charleston Post*, 97
Charlotte, 108
Chater, Melville, 102
Cherokee Indians, 41, 48, 52
Chesapeake Bay, 63
Cheshire, 87
Chester White, 65
Chicago, 71, 75
Chickamauga, 72–74
China, 16, 21, 22, 23, 24, 25, 65, 124
Chincoteague, 63, 64, 93, 100, 101, 115, 116
Chincoteague Bay, 64
Chincoteague National Wildlife Refuge, 115
Chincoteague Wildlife Refuge, 117
Choctaw Indians, 45, 181, 187
Chowan Indians, 40
Cincinnati, 73
Civil War, 10, 78, 79
Clemson, 87
Clifton Forge, 110
climate, 28, 152, 174
Clinton, Bill, 119
CNN, 128
Collier, Charles, 106
Colonial Williamsburg, 131
Columbia, 97, 108, 110
Columbus, 108
Columbus (Georgia), 98
Columbus, Christopher, 29, 32, 52
Comte de Castelnau, Francis de La Porte, 63
Conestogas, 59

## Index

Confederate States of America, 72, 75, 76, 79
Confederate Subsistence Bureau, 74
Coppinger, Lorna, 17
Coppinger, Raymond, 17
corn, 81
Corolla, 119, 120
Corolla Wild Horse Fund, 120
cotton, 67, 71, 72, 73, 79, 80, 99
cowboys, 38
cracker horses, 121
Creath, W.L., 109
Credle, Mary Farrow, 102
Creek Indians, 41, 45, 53, 181
CRISPR-Cas9, 115, 124, 132
Cronon, William, 5
Crosby, Alfred, 3
Crossville (Tennessee), 128
Cuba, 60
Cumberland Gap, 62, 72
Cumberland Island, 44, 95, 102, 103, 120
Cumberland Island National Seashore, 120
Currituck County (North Carolina), 120

*Daily Enquirer* (Columbus, Ga.), 96
Dale, Thomas, 32
Dalmatians, 57
Dan River Kennel Club, 110
Daniels, Pete, 100
Danube River, 20, 23
Darwin, Charles, 2, 15
Daytona Beach, 96
De Soto, Hernando, 30
de Warville, Jacques Pierre Brissot, 170
deer, 40, 45
Delaware, 64
Delmarva Peninsula, 64
Denisovans, 14, 16
dingoes, 166
Djerassi, Carl, 133
Doeg Indians, 36
dogs, 5, 8, 9, 10, 11, 15–20, 24, 25, 27, 28–29, 30, 31–32, 33, 40, 42–43, 45–46, 48, 50, 51, 52, 53–59, 78, 103–11, 112, 131–36, 138, 140
Don Quixote, 66
droving, 60–62, 71–72
Ducos, Pierre, 17
Dumbarton, 136

Duplin County (North Carolina), 124
Durand of Dauphine, 49
Duroc Jersey, 65

earmarks, 34, 37, 45, 66, 67
East Indies, 65
Eastern Shore (Virginia), 100
Ebenezer, 43, 44
Eby, Henry Harrison, 74
*Ecology of Invasions by Plants and Animals*, 126
Egypt, 23
elephants, 130
Ellenton, 136
Elton, Charles, 126
England, 32, 66
English, 11, 30, 33, 37, 39, 42, 49
Ethiopia, 13
Eurasia, 5, 14, 15, 17, 20, 22, 23, 24, 25, 29
Europe, 14, 16, 20, 21, 23, 24, 25
Evans, J.C., 107
*Evening Post* (Charleston), 96
Everglades, 103, 127
evolutionary history (defined), 4
Ewen, William, 44

*Farmer's Register*, 58, 63, 64, 69, 70
Federal Bureau of Fisheries, 102
Federal Relief Administration, 103
fences, 32, 34, 35, 36, 37, 38, 39, 45, 66, 67, 69, 70, 82, 98, 116, 118,
  *See* open range
Fertile Crescent, 20
*Field & Stream*, 89, 130
Field, Marshall, 91
fight-or-flight response, 19
Finch, Atticus, 107
Florida, 6, 8, 27, 30, 45, 46–49, 52, 53, 59, 63, 71, 79, 85, 87, 89, 90, 96, 103, 104, 114, 121, 125, 128, 132
Folsom, Montgomery, 91, 95
*Food, Inc*, 124
*Forest and Stream*, 87, 89, 90, 95, 102, 108, 125, 127
Fort Caroline, 172
Fort Myers, 103
foxes, 18–19
France, 21
French and Indian War, 46
French Broad River, 72
French hounds, 57

frontier, 5, 8–9
Furry, Jean Pierre, 41

Gainesville, 121
Gates, Thomas, 32
genetic erosion, 124
genetics, 15, 16, 21, 22, 25, 27, 29, 115, 130, 136
Georgia, 6, 8, 43–46, 56, 57, 59, 64, 67, 71, 72, 78, 84, 90, 91, 95, 97, 98, 120, 123, 125, 128, 130
Germany, 45
Gilbert, Karl T., 118
Gisolfi, Monica, 123
Gold Coast, 38
Gordon, George Henry, 74
Grant, Ulysses S., 77
Gratiot, Charles, 62
Grayson Highlands State Park, 121
Great Dismal Swamp, 54, 69, 188
Great Smoky Mountains, 92, 137
Greene, Ann Norton, 62
Greensboro, 109
Greenville (South Carolina), 60
Greer, Allan, 9
Gregory, Tappan, 170
Grenville, Richard, 173
greyhounds, 33, 104, 132, 166
Griscom, Lloyd G., 92
Gulf of Mexico, 8
Gullah, 91

Haines, Francis, 30
Hall, Randall L., 97
Hammond, James Henry, 56
Hammond, John, 34
Hamor, Ralph, 32
Harriot, Thomas, 30, 32, 43
Harris, J. William, 90
*Hartford Daily Courant*, 64
Hatteras Inlet, 117
Hawaii, 6
Heathsville, 87
Henrico County (Virginia), 111
Henry, Marguerite, 115, 117
Herodotus, 114
Higginson, Thomas Wentworth, 54
Hilliard, Sam B., 66, 67
Hispaniola, 32, 52
Hog Island, 32
Hoge, Tobe, 88

Hogzilla, 128
Holmes, Thompson, 64
Hooper Bald, 92, 128
*Horse Hoeing Husbandry*, 44
horse slaughter, 115
horses, 5, 8, 9, 10, 11, 20, 22–24, 25, 26, 29, 30, 32, 33, 34–35, 36, 37, 40, 41–42, 44, 46, 48, 50, 53, 59, 62–65, 69, 75–78, 79, 92–95, 97–103, 111, 113–21, 137, 139–40
House of Burgesses, 33, 34, 35, 36, 41, 49
Howard, Marvin, 117
Hubbard, Leonidas, 88, 92
Hungary, 22
hunting, 14, 17, 28, 37, 55, 88–92, 111, 127–30, 140
Huntington, Ellsworth, 99
Hyde, Edward, 39

Iberian Peninsula, 22, 23, 24
Illinois, 74, 80, 81
immuno-contraception, 116
improvement, 66, 81
India, 21
interstate highway system, 113
invasive species, 2, 125, 146
Iowa, 80, 81
Iran, 16, 23
Ireland, 49
Israel, 13
Italy, 92
Ivey, R. DeWitt, 103

Jamaica, 38
James River, 31, 32, 34
James, C.C., 107
Jamestown, 30–33, 42, 50
Jarngain, Milton P., 99
Jefferson, Thomas, 2, 57, 170
Jekyll Island, 44, 91, 95
Jones, Hugh, 49
Jordan, J.L., 110
Judas pig, 127

Kazakhstan, 22, 25
Keiper, Ron, 116
Keith, B. F., 85
Kentucky, 59, 60, 62, 71, 72
Kentucky Derby, 114
Kimmer, Edward, 44
King Victor Emmanuel III, 92

Kirby, Jack Temple, 49, 69, 100, 132
Kirkpatrick, Jay, 116

Lafayette, Marquis de, 57, 60
Lamarck, Jean Baptiste, 15
Lamb, Robert Byron, 62
Larson, Greger, 132
Laurentide ice sheet, 14
Lawson, John, 40, 43
Lee, Harper, 107
Lee, Robert E., 76
Levine, Marsha, 23
Lewis, C. S., 13
Little Horse Island, 121
Little Ice Age, 174
Lost Creek Plantation, 129
Lowcountry (South Carolina), 67, 69, 90, 121, 169
Lutherans, 43

Macon, 58, 96, 97, 110
*Macon Telegraph*, 59, 96
Maghreb, 23
Majewski, John, 9
Marion County (Florida), 114
Marion, Francis, 121
maroon populations, 68, 69
marsh tackys, 95, 121
Martin, John, 176
Maryland, 116
mastiffs, 33, 104, 166
Mayer, Jack, 125
McCarthy, Cormac, 27
McKnight, Tom, 126
*Meat and Livestock Digest*, 87
Mecklenberg County (North Carolina), 98
Mediterranean Sea, 20
megafauna, 27
Melville, Elinor, 3
Mexico, 29, 30
Miami, 127, 134, 135
Miami-Dade Animal Services, 134, 135
Mickle, Walter F., 89
Midwest, 59, 65, 71, 75, 81, 88, 90, 91, 93, 95, 104, 112, 122
Milanich, Jerald, 169
Minor, Peter, 57
Mississippi Era, 28
Mississippi River, 172
Missouri, 72
*Misty of Chincoteague*, 115

Mongolia, 16, 121
Monster Pig, 128–30
Montagu, Ashley, 17
Moore County (North Carolina), 85
Moore, George Gordon, 92
Morgan horses, 59
Morgan, J. P., 91
Morse, Thomas W., 119
Morton, William, 58
Moulton, Horace, 53
Mount Vernon, 57, 60
mules, 10, 59–62, 69, 76, 79, 99–100, 111, 113
Murphy, J. M., 88
mustangs, 30, 62, 63, 93, 103, 117, 120

Narváez, Pánfilo de, 171
Nash, Roderick, 5, 19
National Feral Swine Damage Management Program, 127
*National Geographic*, 102, 128
National Parks Service, 116, 119, 120
*National Stockman and Farmer*, 88
National Zoological Park, 170
Neanderthals, 14, 16
Near East, 16, 21, 23, 25
Needles, 114
Neolithic Revolution, 20, 25
Neuse River, 39
New Deal, 103
New York, 37, 59, 75, 92, 93, 108, 121
*New York Times*, 90, 93, 96
Newenham, Edward, 57
Newfoundland, 57, 58
Newport, Christopher, 31
Norfolk, 108
North Carolina, 6, 8, 30, 38–39, 40, 41, 42, 55, 59, 64, 67, 69, 71, 72, 84, 87, 92, 93, 98, 101, 117, 122, 123–24, 125, 126, 128, 137
Northeast, 59, 65, 70, 75, 88, 90, 91, 95, 104, 108, 111, 112
Northrop, Lucius Bellinger, 74
Northumberland County (Virginia), 87
Nottoway Indians, 39

Ocracoke, 117, 119, 120
*Official Horse Breeds Standards Guide*, 120
Oglethorpe, James, 44
Ohio, 59, 66, 71
Ohio River, 75

Olds, Fred, 101, 102, 119
Olmsted, Frederick Law, 54, 60, 66, 67
Omo River, 13
open range, 9, 10, 34, 35, 41, 45, 50, 62, 66, 67, 69–70, 79, 81–87, 90, 111, 125, *See* fences
Ormond Beach, 96
Ossabaw Island, 130
Ossabaw pigs, 130–31
Outer Banks, 30, 64, 65, 93, 94, 101, 102, 117–20
*Outing Magazine*, 88
Outram, Alan, 22

Palmer, T. S., 125
pannage, 21, 24
Parker, Ruth, 103
Paynes Prairie, 103, 121
Pennsylvania, 49, 59, 95
Pensacola, 45
People for the Ethical Treatment of Animals, 218
Percy, George, 30
Perkins, Samuel, 44
Philadelphia, 75, 121
Philippines, 21
pigs, 5, 8, 9, 10, 11, 20–22, 24, 25, 29, 30, 32, 33, 34, 35, 36, 37, 39, 40–41, 43, 50, 52–53, 65–75, 79, 80–92, 111, 122–31, 137, 140
Pinehurst, 85, 108
Pleistocene, 15, 17, 22, 24, 27, 28, 136
Poinsett, Joel, 53
pointers, 58, 104
Poland-China, 65, 81, 87
Pollan, Michael, 4
Pollock, Thomas, 39
Ponce de Leon, Juan, 171
pony penning, 64, 93, 94, 100, 101, 115, 117
pony swim. *See* pony penning
Porcher, William M., 84
pork, 71, 73, 74, 75, 79, 81, 125
Porter, Katherine Anne, 113
Portugal, 21
Potomac River, 6, 8, 36, 49, 52, 71
Powhatan, 31, 32
Powhatan Indians, 32, 45
Preakness Stakes, 114
Prince William County (Virginia), 70
Przewalski, Noklai Michailovich, 121
Przewalski's horses, 121, *See* wild horses
Puerto Rico, 173

pugs, 104
Pulitzer, Joseph, 91

rabies, 107
race, 14–15
racing
  automobiles, 96–97
  horses, 42, 59, 62, 114–15
radio-telemetry, 127
railroads, 71, 73, 75, 79, 81, 90, 92, 93, 97
Raleigh, 111, 123
Raleigh, Sir Walter, 95
razorbacks, 71, 85, 87, 88, 90
Reconstruction, 84
Red Sea, 13
refrigeration, 75
Reid, Charles Sloan, 88
Revolutionary War, 121
rewilding, 146
Rhode Island, 37
Ribault, Jean, 172
rice, 38, 67, 90
Richmond, 57, 58, 60, 74, 104, 108, 109
Richmond County (Georgia), 85
Ricketson, Margaret Carnegie, 103
Ricketson, Oliver, 103
Ringwalt, Samuel, 75
Ritvo, Harriet, xi, 3
Rocky Mount (North Carolina), 123
Roman Empire, 24
Romans, Bernard, 45
Rose, Carl G., 114
Rosinante, 66
Rowlsley, William, 33
Royal Gift, 60
Ruffin, Edmund, 64, 69–70
Russell, Edmund, 4, 17, 130

Saint-Exupéry, Antoine de, 139
Salzburgers, 43, 44
Sampson County (North Carolina), 124
Santee Hunting Club, 91
Santee River, 171
Sarasota, 169
Savannah, 43, 46, 49, 53, 56, 58, 96, 97
Savannah River, 43, 169, 172
Savannah River Ecology Lab, 135
Savannah River Site, 135, 136
Scarburgh, Edmund, 36
Schoepf, Johann David, 60, 65
Schwartz, Marion, 135
Scotland, 49

# Index

*Scribner's*, 93
*Sea Venture*, 32
Seaboard Air Line Railway, 90
Secretariat, 114
Seminole Indians, 46–48, 53, 103
Seminole Wars, 63
setters, 104
Shackleford Banks, 102, 119, 120
Shakespeare, William, 174
sharecropping, 80
shatter zone, 9
Shay, William, 123
sheep, 57, 58
sheepdogs, 57
Shenandoah Valley, 49
Shepard, S. M., 88
Sherman, William T., 105
Shields, R. L., 87
Sholtz, David, 103
Shuanghui International, 124
Siberia, 14, 16, 18, 27, 28
Silk Road, 23
Sinai Peninsula, 13
Skinner, John, 62
slavery, 38, 40, 41, 42, 49–50, 53–56, 67–69, 71
smallpox, 39
Smith, John, 31, 43, 52
Smithfield Foods, 124, 140
Smyth, J. F. D., 42
Snyder, Henry, 91
South (defined), 6–8
South Carolina, 6, 8, 37–38, 40, 41, 42, 46, 50, 56, 59, 60, 62, 63, 67, 69, 70, 71, 72, 74, 78, 84, 87, 90, 97, 121, 128, 135
*Southern Agriculturalist*, 58
*Southern Cultivator*, 56, 58, 72
southern exceptionalism, 150
*Southern Planter*, 57, 70
*Southern Pork Production*, 81
Spain, 58, 60
spaniels, 57
Spanish, 29, 33, 38, 46, 171
Spartanburg, 110
Spotswood, Alexander, 39
St. Augustine, 46, 65
St. Bernard, 104
St. Johns River, 46, 48, 90, 169, 172
Starving Time, 32
Stegeman, LeRoy, 92, 125
Steinberg, Ted, 75

Stewart, Mart, 9
Stone, Jamison, 128
*Stormy*, 117
Strait of Gibraltar, 23
Sun Belt, 6
*Sunny South*, 108
Swine Development Center, 123

Tampa, 109
*Tampa Tribune*, 109
Tar Heel (North Carolina), 124
Taylor, John, 69
Taylor, Zachary, 53
Tchakerin, Viken, 9
Tennessee, 59, 60, 71, 72, 74, 75, 76, 128
terriers, 57, 107
Texas, 76, 137
*The State* (Columbia), 97
*The Tempest*, 174
Thomas, Carrie, 106
Thomas, D. W., 110
thoroughbreds, 114
tobacco, 123
total-confinement agriculture, 123–24
Townsend, W. R., 91
tractors, 99, 100
Tull, Jethro, 44
Turkey, 16, 20
Turner, Frederick Jackson, 8
Turner, Nat, 188
Tuscarora Indians, 39
Tuscarora War, 40

US Army Corp of Engineers, 62
US Fish and Wildlife Service, 137
underground economy, 67
United Kennel Club, 136
United States Forestry Service, 102
Upton, Thomas, 44
USDA, 81, 100, 114, 115, 127

Van Ewing, Perry, 81
Vanderbilt, William K., 91
Vermont, 59
Vienna (Georgia), 110
Virginia, 6, 8, 29, 30–37, 41, 42, 49, 50, 52, 54, 57, 59, 60, 66, 69, 70, 71, 74, 76, 78, 79, 93, 100, 116, 121, 124, 131, 137
Virginia Peninsula, 34
Vonnegut, Kurt, 80

Wallace, Alfred Russell, 2
Ward, A. J., 103
Washington (state), 137
Washington, D. C., 121
Washington, George, 56, 57, 60
Weir, Addison M., 105
White, John, 30
White, Samuel, 24
Whitson, William, 91
wild boars, 20, 21, 92, 128
wild horses, 22, 23, 121, *See* Przewalski's horses
wilderness/wildness, 5, 19, 139, 147
Wilkes County (North Carolina), 110
Williams, Edward, 34, 42
Williamsburg, 49
Wilmington, 85, 110

wolfhounds, 57, 166
wolves, 15, 16, 18, 19, 25, 29, 35, 48–49, 136–37, 170, 176
Woodland Era, 28, 29
Woodson, Jack, 101
World War II, 132

Yamassee Indians, 40
Yamassee War, 40, 43
Yangtze River, 16
Ybor City, 109
Yeats, William Butler, 1
York River, 34, 36
Young, Norwood, 119

Zeder, Melinda, 17

Other Books in the Series (*continued from p.iii*)

Matthew D. Evenden *Fish versus Power: An Environmental History of the Fraser River*
Nancy J. Jacobs *Environment, Power, and Injustice: A South African History*
Adam Rome *The Bulldozer in the Countryside: Suburban Sprawl and the Rise of American Environmentalism*
Judith Shapiro *Mao's War Against Nature: Politics and the Environment in Revolutionary China*
Edmund Russell *War and Nature: Fighting Humans and Insects with Chemicals from World War I to Silent Spring*
Andrew Isenberg *The Destruction of the Bison: An Environmental History*
Thomas Dunlap *Nature and the English Diaspora*
Robert B. Marks *Tigers, Rice, Silk, and Silt: Environment and Economy in Late Imperial South China*
Mark Elvin and Tsui'jung Liu *Sediments of Time: Environment and Society in Chinese History*
Richard H. Grove *Green Imperialism: Colonial Expansion, Tropical Island Edens and the Origins of Environmentalism, 1600–1860*
Elinor G. K. Melville *A Plague of Sheep: Environmental Consequences of the Conquest of Mexico*
J. R. McNeill *The Mountains of the Mediterranean World: An Environmental History*
Theodore Steinberg *Nature Incorporated: Industrialization and the Waters of New England*
Timothy Silver *A New Face on the Countryside: Indians, Colonists, and Slaves in the South Atlantic Forests, 1500–1800*
Michael Williams *Americans and Their Forests: A Historical Geography*
Donald Worster *The Ends of the Earth: Perspectives on Modern Environmental History*
Samuel P. Hays *Beauty, Health, and Permanence: Environmental Politics in the United States, 1955–1985*
Warren Dean *Brazil and the Struggle for Rubber: A Study in Environmental History*
Robert Harms *Games Against Nature: An Eco-Cultural History of the Nunu of Equatorial Africa*
Arthur F. McEvoy *The Fisherman's Problem: Ecology and Law in the California Fisheries, 1850–1980*
Alfred W. Crosby *Ecological Imperialism: The Biological Expansion of Europe, 900–1900, Second Edition*
Kenneth F. Kiple *The Caribbean Slave: A Biological History*
Donald Worster *Nature's Economy: A History of Ecological Ideas, Second Edition*

Lightning Source UK Ltd.
Milton Keynes UK
UKHW011249180621
385743UK00001B/43